Dick PeR

$60.95

Dick Pepin

FDDI: A High Speed Network

Amit Shah

G. Ramakrishnan

PTR Prentice Hall
Englewood Cliffs, New Jersey 07632

Library of Congress Cataloging-in-Publication Data

```
Shah, Amit.
     FDDI: a high speed network / Amit Shah, G. Ramakrishnan.
       p.  cm,
     Includes bibliographical references and index.
     ISBN 0-13-308388-8
     1. Local area networks (Computer networks) 2. Fiber Distributed
     Data Interface (Computer network standard) I. Ramakrishnan, G.
     II. Title.
TK5105.7.S48  1994                                         93-11373
004.6'8--dc20                                                  CIP
```

Editorial/production supervision
   and interior design: *Dit and Dominick Mosco*
Cover design: *Jerry Votta*
Manufacturing buyer: *Mary E. McCartney*
Acquisitions editor: *Karen Gettman*

 ©1994 by PTR Prentice-Hall, Inc.
A Simon & Schuster Company
Englewood Cliffs, New Jersey 07632

The publisher offers discounts on this book when ordered in bulk quantities. For more information, contact:

> Corporate Sales Department
> PTR Prentice Hall
> 113 Sylvan Avenue
> Englewood Cliffs, NJ 07632
>
> Phone: 201-592-2863
> Fax: 201-592-2249

ISBN 0-13-308388-8

Printed in the United States of America
10 9 8 7 6 5 4 3 2 1

Prentice-Hall International (UK) Limited, *London*
Prentice-Hall of Australia Pty. Limited, *Sydney*
Prentice-Hall Canada Inc., *Toronto*
Prentice-Hall Hispanoamericana, S.A., *Mexico*
Prentice-Hall of India Private Limited, *New Delhi*
Prentice-Hall of Japan, Inc., *Tokyo*
Simon & Schuster Asia Pte. Ltd., *Singapore*
Editora Prentice-Hall do Brasil, Ltda., *Rio de Janeiro*

Contents

Preface

Welcome to this book! After teaching FDDI and other networking technologies to a wide assortment of people including design engineers, MIS managers, instructors, sales and marketing persons, field engineers and students, we have come to appreciate the different perspective each user has to the same technology. In this book, we have attempted to look at the network while considering these perspectives along with those of the designers of FDDI, (ourselves), and balance them with the systems aspects of FDDI to present a book which is neither too technical nor too simplistic.

Questions vary from, "How should I use it?" to "Why was this done, and not that?". Here, we try to clarify the implementation details and hard questions an end user should ask the vendor when buying FDDI equipment. In addition, a glossary is provided to explain some of the acronyms, terms and phrases floating around the networking community in general and especially the FDDI community.

Because of our involvement in the development of the FDDI family of standards, it has often been impossible to present the information in an entirely neutral manner. We hope our personal biases offer some insight into the often inscrutable text of the standards documents, and contribute to a book in which the authors understand the subject and explain it simply and succinctly.

After extensive revision and the addition of several illustrations to enhance the textual explanations, we are confident of this book's technical content. However, given the in-depth treatment of a complex subject, it is possible some technical error might have escaped us even after all the scrutiny.

We would like to thank numerous individuals who directly or indirectly contributed to the writing of this book. First of all, we would like to thank Karen Gettman and Dit Mosco of Prentice Hall without whom the book would not have happened. Next, our thanks to all our numerous reviewers. We must especially thank Scott Smith of the University of Colorado, Jayant Kadambi of AT&T, and Gene Gonzales of AMD for their many helpful suggestions, most of which were incorporated into the book. No missing comma, misconstrued sentence, illogical statement or unreasonable explanation ever made it past Robert DeMong of AMD. For that, we are indebted as he made our writing much clearer and more succinct.

Finally, without the support of our families, this book would never have seen the light of day. Thank you Bela and Jayashri and Deepa for your encouragement and steadfast support.

About This Book

1.1 INTRODUCTION

Every office, big or small is equipped with a desktop computer. Personal computers have become ubiquitous in every day life. By and large, these desktop computers have not been inter-connected. The past decade has seen an increasing interest in computer communications. In fact, computer communications is an accepted, necessary evil today. The computer communications market is growing at a rapid pace and has figured prominently in research and development over the past ten years. Computer networking is vital to businesses and is touted as the next technology frontier.

Today's networks can be categorized as Local Area Networks (LANs), Metropolitan Area Networks (MANs), and Wide Area Networks (WANs). In the past, the term network was synonymous with what we today call Wide Area Networks or public networks. Networks can range from dial-up modems to gigabit per second national networks. Anything which connects two computers can be called a network. Thus, even the lowly RS-232 is a point-to-point network. On the other hand, the *Internet* is a sophisticated conglomeration of computers and links connecting a large section of the world together.

What is a LAN? This question has been asked often and answered differently. To some, the difference lies in the controlling authority: users (LANs) versus Post Telephone and Telegraph (PTT) authorities for WANs. To others, it is a matter of the geographical spread. A LAN is traditionally within a floor, building or campus. A WAN connects across cities and countries. However, the key difference lies in the network protocols for a LAN and a WAN. A LAN provides a guaranteed upper-bound on the transmission time and generally provides significantly lower error rates. Considering that LANs were developed at a time when memory was expensive, this meant that LAN interface units require minimum buffering.

When were LANs developed? The first LANs were proposed around the late 60s and early 70s. The first commercial products were the Attached Resource Computer (ARC) from Datapoint Corporations and the SILK (System für Integrierte Lokale Kommunikation). These were followed by the first LAN systems which had a dramatic effect on the LAN market: DIX (Digital Intel Xerox) Ethernet and the Cambridge Ring. This was followed by the IBM Token-Ring. These networks ranged in speed from 1 Mbps to 16 Mbps. Around 1982, a significantly higher speed network was proposed which would integrate data, voice and video on a single network. Thus was born FDDI or Fiber Distributed Data Interface.

1.2 THE NEED FOR HIGH SPEED LANS

This is a subject which has been extensively discussed throughout the development of the LAN market. In the early 1980s, 3 Mbps Ethernet was considered as sufficient for the next decade. Ten Mbps Ethernet has become the desktop connection of choice within a few years of its introduction.

The desktop computer has become increasingly powerful over the past ten years, and several technological trends indicate that at the current rate of development, the network pipeline will soon become the bottleneck in the end-to-end system connection.

The increasing CPU power has led to the development of several distributed applications such as client-server computing, multimedia computing and distributed databases. These applications require high *burst* and *aggregate* bandwidth.

Burst bandwidth is the instantaneous capacity required of a network by a system. This load may not occur constantly. An example of this is an imaging application where large images are exchanged over the network periodically. The faster the network, the higher the burst bandwidth of a network.

Aggregate bandwidth is the cumulative bandwidth required by each node and is independent of the burst bandwidth. An example of this is multiple simultaneous voice sessions. Each session does not require large bandwidth but the cumulative bandwidth can be fairly large. Higher aggregate bandwidth does not imply a faster network. A backbone network requires higher aggregate bandwidth.

Ethernet and Token-Ring networks cannot provide the burst bandwidth today. These networks are limited in their ability to provide aggregate bandwidth by the number of nodes that they can support.

1.3 NEW APPLICATIONS

Traditional computer applications have included printer-sharing and file-sharing. These applications were typically restricted to a small workgroup within an organization and the computer networks grew adhoc as and when the need to share printers and/or files arose. With the advent of electronic mail (e-mail), computer networking has become enterprise-wide, and corporate-wide connectivity has become the norm. Thereafter, the client-server paradigm has become prevalent with centralized departmental, divisional and corporate servers and hundreds of client workstations. This has often imposed bandwidth bottlenecks leading to the need for higher speed backbone networks.

On the engineering workstation front, the increasing popularity of networked CAD/CAM applications has swamped existing networks. This, coupled with the imminent arrival of multimedia applications such as desktop videoconferencing, distance learning and distributed computing has led to the development of new higher speed networks such as Fiber Distributed Data Interface (FDDI).

1.4 ABOUT THIS BOOK

This book explains the details of FDDI from two perspectives: the engineer's and the user's. The first part of the book is suitable for the design engineer, project manager, graduate student or MIS manager desiring to gain a perspective on the technology.

The first eight chapters explain the history behind FDDI, the standards effort, the different parts of the FDDI set of specifications, the network topologies and the network management aspects of FDDI.

The second part of the book (chapters nine through eleven) is about the system implementations of FDDI, the internetworking issues of FDDI, the performance issues, cabling and troubleshooting.

Chapter two explains the fundamentals of local area networks and provides a comparison of FDDI with other popular local area networks such as Ethernet and Token-Ring.

Chapter three provides an overview of FDDI explaining the concepts of network operation and topology.

Chapter four discusses the media access control mechanism for FDDI with numerous examples on the operation of timers and counters. Address recognition, ring initialization, ring operation and advanced issues such as fault recovery are described.

Chapter five discusses the physical layer portion of FDDI and the functions therein.

Chapter six is the transmission characteristics of FDDI including transceivers, connectors and cable specifications.

Chapters seven and eight discuss the station management specification of FDDI. The local station management is discussed in chapter seven and the remote management is discussed in chapter eight. The use of the Simple Network Management protocol to manage FDDI networks is also described.

Chapter nine discusses the integration of FDDI in a heterogeneous environment with existing networks such as Ethernet and Token-Ring. The principles of MAC-level bridging and network layer routing are described. Internetworking devices such as bridges and routers are discussed. Suggestions on selecting bridges and routers are provided to help the MIS manager in making an intelligent decision.

Chapter ten talks about the medium itself: the wire. The impact on existing cabling, installing new cabling, planning for the future and trade-offs are some of the issues discussed in this chapter. The implications of FDDI over copper are also indicated.

Chapter eleven talks about the performance issues in FDDI. Various performance parameters such as throughput, latency, and efficiency are discussed. Guidelines on setting of the ring parameters are provided.

Chapter twelve provides the reader with important trouble-shooting advice. This advice is for the user as well as the design engineer. Lessons learned from interoperability testing are discussed.

Last, a glossary of terms used in the book is provided. The reader is encouraged to refer to the glossary as much as possible for acronyms which may not have been expanded in the text.

FDDI Standards and Basics

2.1 INTRODUCTION

This chapter briefly explores the basics of FDDI. It presents an overview of the set of FDDI standards and a comparison with other networks.

FDDI stands for Fiber Distributed Data Interface. It is a 100 Mbps data rate local area network standard. The topology is a point-to-point connection of links connected in a logical ring. There are two such rings with one ring configured as a backup. Each station connects to both rings. There can be 500 such stations on the network with a maximum ring-size of 200 km[1]. The maximum distance between each node is 2 km[2].

According to the Open Systems Interconnection Reference Model (OSI-RM), FDDI specifies layer 1 (physical layer) and part of layer 2 (data link layer). The data link layer is responsible for maintaining the integrity of information exchanged between two points. In a LAN environment, because most LANs are *shared media* networks, the layer 6 is subdivided into two parts: logical link control (LLC) and media access control (MAC). The LLC is the same for different *subnetworks*. Examples of subnetworks are IEEE 802.3 (CSMA/CD, very similar to Ethernet), IEEE 802.5 (Token-Ring), and ANSI FDDI amongst others. Different LAN subnetworks have different mechanisms for accessing the shared media. The access mechanisms are specified in the media access control (MAC) specification of the LANs. The layer 1 FDDI specification consists of two parts: physical media dependent (PMD) and PHY. The PHY is independent of the media which can be multimode fiber, single mode fiber, shielded twisted pair, and so on.

The FDDI specifications have been developed by the American National Standards Institute (ANSI). Within ANSI, the X3T9 committee is chartered with the development of high-speed I/O interfaces such as FDDI, Fiber Channel and HIPPI. The X3T9 technical committee responsible for the development of FDDI is X3T9.5. The development of the FDDI standards officially began in June 1983 although prior work had begun in the 1981 time-frame at various companies such as Sperry and Burroughs. The first FDDI standard, the FDDI MAC, became available in 1987 (X3.139-1987). Thereafter, the FDDI PHY became an ANSI standard in 1988 (X3.148-1988). This was followed by the FDDI PMD (multimode fiber), FDDI SMF-PMD, and SMT. The

---

1. These are default values. The standard allows for different values.

2. For single-mode fiber, the distance can be up to 40 km.

FDDI MAC, PHY and PMD have become International Standards (IS 9314-2, 9314-1, 9314-3 respectively).

A standard is not frozen when a project becomes an ANSI or ISO standard. The standard is put in a revision mode for five years. At the end of five years, the viability of the standard is reexamined. If it is determined that the target market and technology for the standard still apply, the committee votes to extend its life-cycle. This process is called a reaffirmation of the standard. The FDDI PHY and MAC specifications were recently (1991) reaffirmed for another five-year period.

2.2 PHYSICAL MEDIA DEPENDENT SUBLAYER (PMD)

The PMD defines the type of media interconnection and its characteristics such as transmitter power, frequencies, receiver sensitivities, and so on. It also specifies the maximum repeater-less distance between two nodes. The first media targeted for FDDI was fiber. There are various kinds of fiber and the committee standardized the 62.5/125 micron multimode fiber. The transceivers required for multimode fiber are cheaper than the single mode fiber transceivers (used in telecommunications networks) and provides high bandwidth and low attenuation up to a distance of 2 km. However, 62.5 micron fiber characteristics restrict its use in long distance communications.

In order to accommodate inter-nodal distances greater than a few kilometers to allow FDDI to be used between campuses and/or buildings and make use of existing fiber in the public domain, ANSI X3T9.5 approved a 50 micron Single Mode Fiber (SMF) as an alternative media. The 50 micron SMF can span distances up to 40 km (repeater-less). This requires the use of powerful transmitters such as laser diodes.

The biggest difference between multimode fiber and single mode fiber is the cost. Multimode fiber can work with lower cost LEDs. The single mode fiber requires expensive lasers and receivers.

Currently, work is under way to develop a low-cost alternative to the 1,500 nm multimode and single mode fiber. Several low-cost alternatives were proposed: plastic fiber, 850 nm fiber, shielded twisted pair, unshielded twisted pair, and DIW twisted pair. The desire to bring low-cost FDDI to the desk-top prompted the committee to set up two working groups—one dealing with low-cost fiber alternatives and the other dealing with FDDI over copper cabling. The low-cost fiber working group has completed development of a low-cost fiber PMD specification which uses the same multimode fiber but restricts the maximum distance to 500 meters. This allows cheaper transceivers to be used.

2.3 PHYSICAL SUBLAYER (PHY)

The function of the PHY is to synchronize the clock with the incoming signal provided by the PMD, encode and decode data and control symbols, interface with the higher layers and initialization of the medium (fiber, copper). The FDDI PHY is different from the PHY specification of Ethernet. It bears more similarity to Token-Ring than Ethernet. The PHY protocol is embedded in VLSI circuits.

2.4 MEDIA ACCESS CONTROL (MAC)

The function of the MAC is to provide ordered (fair) access to the medium, addressing capabilities, data checking (frame check sequence generation and verification), packaging of data (framing) for transportation over the network, removal of incoming traffic (own frames), and repetition of traffic (frames). The FDDI MAC also specifies a class of control frames called MAC frames which are used to execute low-level (MAC-level) protocols such as ring initialization (claim process) and fault isolation (beacon process). The MAC protocol is embedded in silicon.

2.5 STATION MANAGEMENT (SMT)

This is not an OSI-RM specification. It is a management entity which is very closely coupled with the FDDI MAC, PHY and PMD. This management entity provides both high-level (network entities) and low-level (MAC, PHY, PMD entities) monitoring and management capabilities to the network manager. SMT consists of frame-based management (network-level), ring management (node-level), and connection management (link-level).

Frame-based management provides for remote management and monitoring of stations on an FDDI network. A class of frames are defined to support SMT frame-based protocols. This alleviates the need for protocol dependent network management tools such as Simple Network Management protocol (SNMP) and Common Management Interface protocol (CMIP). Frame-based management is currently not implemented in silicon but in software.

Connection management and ring management deal with the initialization of the node, inserting on to the ring and incorporating the MAC into the token-path. Several of the low-level functions of connection management have been incorporated into VLSI solutions.

2.6 COMPARISON WITH OTHER NETWORKS

Most of the popular networks today are shared media LANs. FDDI is also a shared media LAN. It was developed to serve as an integrated voice/video/data network. Table 2.1 provides a comparison of FDDI with two of the most popular LANs–Ethernet and Token-Ring.

As can be seen from the table, FDDI is very similar to Token-Ring. A lot of FDDI features have been derived from Token-Ring.

Table 2.1 Comparison of FDDI, Ethernet and Token-Ring

Features	FDDI	Ethernet	Token-Ring
Raw transmission rate	125 Mbaud	20 Mbaud	8 & 32 Mbaud
Raw data rate	100 Mbps	10 Mbps	4 & 16 Mbps
Media[2]	Fiber, STP, DG-UTP	Coax, UTP, STP, Fiber	STP, DG-UTP, Fiber
Signal Encoding	4B / 5B	Manchester	Differential Manchester
Clocking	Distributed Transmit and Receive clocks	-NA-	Centralized Active Ring monitor
Maximum Frame size (bytes)[3]	4500	1526	4500(4), 18000 (16)
Frame format	SD, FC, DA, SA,INFO, FCS, ED, FS	SD, DA, SA, TYPE, INFO, FCS	SD, AC, FC, DA, SA, INFO, FCS, ED, FS
Priority Levels	Synchronous, 8 levels of Asynchronous	No Priority	8 levels
Priority Handling	Uses Station priority	-NA-	Uses ring priority
Transmission Process	Byte-level manipulation	Bit level manipulation	Bit level manipulation
Access Protocol	Token Passing–Timed Token Rotation protocol	Carrier Sense Multiple Access/Collision Detection (CSMA/CD)	Token Passing

Table 2.1 Comparison of FDDI, Ethernet and Token-Ring (Continued)

Features	FDDI	Ethernet	Token-Ring
Network Transmission	Token is captured by stripping from network. Thereafter station transmits (multiple) frames	Sense Idle (empty) network (wire) and transmit a frame	Token is captured by setting status bit, converting it to frame. Single or multiple frames are transmitted in 4 and 16 Mbps respectively.
Token release	After transmissions complete	-NA- (CSMA/CD mechanism)	After reception of transmitted frames (4 Mbps); after transmissions (16 Mbps)
Maximum Coverage[4]	200 km	2.8 km	configuration dependent
Maximum Nodes[5]	500	1024	260
Maximum Distance between Stations	2 km	2.8 km	300 m WC to station (recommended 100 m)
Topology	Logical Dual - Ring, or Dual-Ring of trees	Bus	Logical single ring
Physical Topology	Ring, Hierarchical star, Star	Star, Bus, Hierarchical Star	Ring, Star
Network Transmission order (Header)	msb first*	msb first*	msb first*
Network Transmission order (Info. field)	msb first	lsb first	msb first

Table 2.1 Comparison of FDDI, Ethernet and Token-Ring (Continued)

Features	FDDI	Ethernet	Token-Ring
SD:: Start Delimiter	FC :: Frame Control	AC :: Access Control	DA :: Destination Address
SA :: Source Address	FCS :: Frame Check Sequence	ED :: Ending Delimiter	FS :: Frame Status
UTP :: Unshielded Twisted Pair	STP :: Shielded Twisted Pair	msb :: most significant bit	

* msb first in the header is actually the individual/group bit of the address field transmitted first

2. This is a list of the typical media used. Any alternative media can be used in any of the networks.

3. This includes 8 bytes of preamble in FDDI and Ethernet.

4. This is different for different media. In FDDI, a copper media network is restricted to a much smaller size. Similarly, in Ethernet the network span varies with the type of media used, for example 185 m for Cheapernet.

5. This is also different for different media. For example, in FDDI, the maximum distance is restricted to 100 meters for copper cabling.

CHAPTER 3

FDDI Nodes and Network Topology

3.1 NETWORK ARCHITECTURE

FDDI specifies a dual ring of trees architecture (figure 3.1). It is a point-to-point network, where each node is daisy-chained to the previous node. There are two such counter-rotating rings. The rule for forming an FDDI network is simple–*No greater than two logical data paths*.

The set of dual rings is referred to as the trunk ring. The dual rings are counter-rotating. This allows for easy reconfiguration in the case of a single fault (figure 3.2). The dual ring architecture refers to a *logical* dual ring. Physically, the cables may be configured as a star. Typically, only one ring is used for data transfer. This ring is called the primary ring. The second ring is used as a backup, to be used in case of a fault. This ring is called the secondary ring.

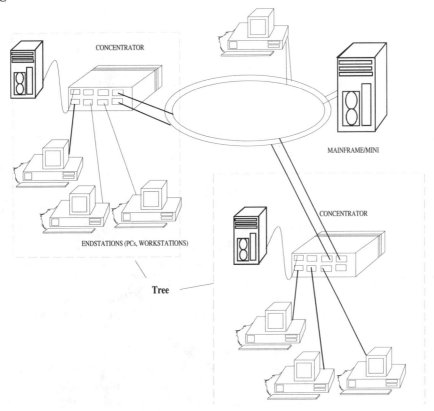

Figure 3.1 Dual ring of trees.

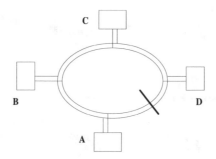

Figure 3.2a   Cable-fault occurs in dual ring between stations A and D.

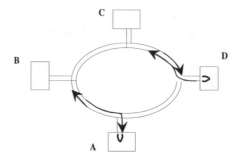

Figure 3.2b   Stations A and D reconfigure around cable break. Double ring becomes a single ring.

3.1.1   Definitions

Trunk ring implies the set of dual rings.

A *node* is a generic term applying to any active element in an FDDI network. A node contains at least one port: PHY + PHY + PMD.

A *station* is defined as a node with a MAC. A station is an addressable entity. For instance, a frame can be sent to a station but not to a repeater.

A *dual attachment station (DAS)* attaches to the trunk ring. A *single attachment station (SAS)* attaches to one ring only. A SAS usually attaches to the trunk ring through a concentrator.

An *optical bypass* or *bypass switch* or simply *bypass* is a device which provides ring continuity in the presence of a node fault as shown in figures 3.3a, b and c.

A *concentrator* is a type of a node which provides connections to single attachment nodes. A concentrator itself may be capable of directly attaching to the trunk ring (*dual attachment concentrator-DAC*), single ring only (*single attachment concentrator-SAC*), or forming the root of a tree (*null attachment concentrator-NAC*).

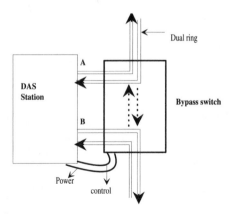

Figure 3.3a Bypass switch or relay.

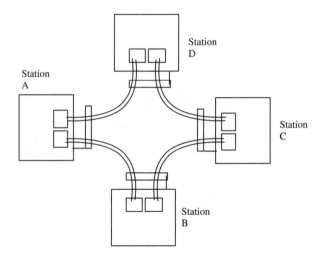

Figure 3.3b Normal dual ring operation.

3.2 NETWORK CONNECTIONS

In order to comply with the rules for forming legal FDDI topologies and to support all the different node configurations possible, FDDI ports are instantiated with certain identities. By exchanging identities during connection initialization, it is possible to ensure that no illegal or undesirable connections are formed. There are four types of ports defined in FDDI SMT: A, B, S and M.

The A port is the port in a DAS (or DAC) through which the secondary ring exits. The B port is the port in a DAS (or DAC) through which the primary ring exits. The A and B ports always occur together. Every station connecting to the trunk has a minimum of both an A and a B port.

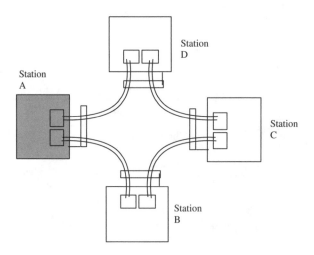

Figure 3.3c   Station A bypassed while maintaining dual ring for stations B, C and D.

M ports are distribution ports found in concentrators. The M port is the port in a DAC, SAC or NAC which extends the primary or secondary ring and allows single connection nodes to be a part of the FDDI network. There are usually greater than two such ports in a concentrator; typically, 12 to 24 ports. Normally, an S port connects to an M port in a concentrator.

The port differentiation is achieved by labeling ports in a DAS, SAS and concentrator differently, ascribing different properties to the ports, and providing different keys to the connectors. Most FDDI systems are shipped with preassigned port types.

The ports of a DAS can only attach in a certain manner to the ports of another DAS (figure 3.4). Thus, an A to A connection is undesirable because it forms a twisted ring (figure 3.5). As can be seen from the figure, both stations, *stupid* and *half-stupid* are connected A to A. An A port is defined as *incoming of the primary and outgoing of the secondary*. Thus, their MACs are placed on what each considers to be the primary path. Because of the A to A connection, the paths are actually swapped (twisted ring) and the *stupid* MAC ends up being on a different path than the *half-stupid* one. This results in the two MACs being unable to communicate with each other. By symmetry, this problem is also applicable to the B ports. This problem can easily be avoided by keying the connectors such that an A to A or B to B connection cannot be formed.

The S port is the SAS (or SAC) port which connects to one ring only. The S port typically connects to an M port. The S ports can be visualized as the leaves of a tree-branch, connecting to the branch via the M ports (figure 3.6a). Most end-stations are expected to be SAS connections.

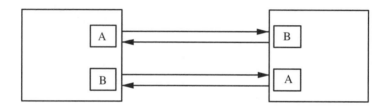

Figure 3.4   A two station DAS network. The normal port connections are A to B and B to A.

Table 3.1 at the end of the chapter explains the reasons why the different connections are legal, illegal, or undesirable. Although the A to A or B to B connections can be easily avoided, there are other connections which are not so easy to avoid or detect. An example of this is an A to M and B to M connection. Such connections are typically taken care of in the SMT software provided by each vendor.

A *tree* connection is formed when an A, B, or S port connects to an M port. In this configuration, the node is not attaching to the trunk ring but to another node's (concentrator) port on a *concentrator tree* (figure 3.1).

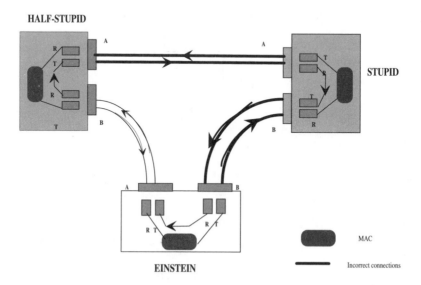

Figure 3.5   Half-stupid and stupid have connected their supposed incoming primary path to their respective A ports. This leads to their MACs being on different paths. It can happen due to incorrect keying of the connectors. This is also called a twisted ring.

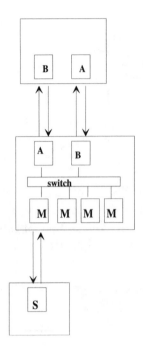

Figure 3.6a   An S port connects to an M port of a concentrator.

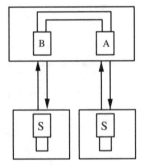

Figure 3.6b   An S port connecting to an A or B port reduces the dual ring to a single ring.

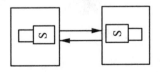

Figure 3.6c   An S port to S port connection always forms a two station single ring.

A *peer* connection is formed when an A, B or S port connects to an A, B, or S port. In a peer configuration, the node is attaching to the trunk ring. The S port connection to A or B ports is a degenerate case which causes the ring to wrap (3.6b). The S port to S port connection is a trivial case of only two stations on a ring when both stations are SAS (figure 3.6c). However, it is possible to see an S to S connection when two concentrator trees connect to each other through SACs. In this case, there is only one logical ring and even a single fault will segment the network.

3.3   FDDI NODES

This section examines the various FDDI nodes. As explained earlier, FDDI nodes can primarily be divided into two kinds: dual attachment and single attachment. Furthermore, there are end stations and concentrators. We examine the various types of end stations and concentrators and their possible applications.

3.3.1   SAS

A SAS node must have a MAC, otherwise it is a repeater. Although repeaters are not explicitly disallowed, their functionality is normally not required in a SAS configuration (figure 3.7).

A SAS contains a fiber or copper MIC, a PHY and a MAC. It is feasible that all of the components are integrated onto the same die in silicon. The SAS also contains the software portions of SMT which are not embedded in silicon. Normally a SAS will not have more than one MAC.

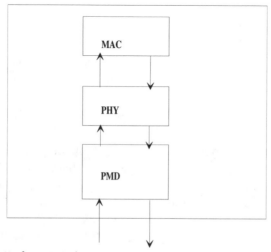

Figure 3.7   Single attachment station.

Most desktop computers will have a SAS connection. A SAS connection is very simple compared to a DAS and can be extremely low-cost. It is thus ideal for desktop connectivity even though a cable fault may isolate the node from the network. Moreover, switching off the desktop at the end of the day will not break the network as the desktop FDDI adapter is connected to a concentrator at the other end in the wiring closet. The concentrator senses that the adapter is shut-off and turns off the port to which the desktop is connected.

A SAS connection is normally preferred under two conditions:

• Low cost connection is desired

• Minimum network reconfiguration due to node insertion/deinsertion is desired

A ring topology is fairly susceptible to cable faults. Most network problems occur because of a loose connection caused by someone stepping on the cable or twisting or pulling on the connection to the computer. A single such fault can cause a physical ring network to crash. FDDI obviates the problem to a certain extent by providing a dual-ring architecture with the second ring primarily intended for back-up purposes. However, multiple faults can cause the FDDI network to segment.

This problem can be circumvented by providing for concentrators on the dual ring. These concentrators can be placed in a safe, isolated place (such as near a patch-panel) to avoid connection problems. All other stations can attach to the concentrator as single attached. Most FDDI networks will have a similar topology. It is even possible for other stations to have two connections (DAS) but not connected to the trunk ring. This configuration is called *dual homing* and is discussed later in this chapter.

3.3.2 DAS

A DAS also by its very definition as a station must have a MAC. Typically, a DAS has two ports, one MAC and one SMT (software, firmware or VLSI) as shown in figure 3.8. As stated earlier, a DAS attaches to a trunk ring. However, it is possible for a DAS to be attached in a concentrator tree.

A DAS by the nature of its dual connections offers more fault tolerance than a SAS. It also offers two possible data paths. If a DAS has two MACs (figure 3.9), it is possible to place a MAC on each of the primary and secondary rings to do simultaneous data transmissions on both rings. This gives a 200 Mbps throughput. One application of this dual ring transmissions would be to use one ring for normal traffic and the other ring for data backups. This would not interrupt the network traffic on the primary ring and the full 100 Mb bandwidth of the secondary ring is used for faster backups. Thus file servers and high speed I/O devices may be dual MAC DAS. This application can be problematic in a tree architecture as SAS stations will have access to one ring only.

A DAS connection is more expensive than a SAS connection. Typically, most DAS connections will be fiber and will be used in back-bone or high-end applications (such as dual-homing for fault tolerance) on the trunk ring. A lower-cost fault-tolerant configuration is possible by using two SAS connections.

What happens if both rings are used for data transfer and a fault occurs? How does it impact the upper layer protocols?

The SMT for a DAS-dual MAC has to ensure that if the rings are merged, the network will remain stable.

Upper-layer protocols provide data to the network interface driver which transmits the data onto the network. There is a special protocol which maintains the database of the mapping of upper-layer addresses of various stations to their physical addresses. This includes the mapping of the host station's upper-layer address to the physical address. This protocol is the Address Resolution Protocol (ARP). ARP database tables provide the translation mechanism to translate an upper-layer destination name (such as amit_shah) to the physical address (the actual mailing address). When a station has two MACs and hence two physical addresses, the ARP has to be modified to ensure stability in the presence of a cable fault which can merge two rings, and split a ring into two separate rings (and hence networks).

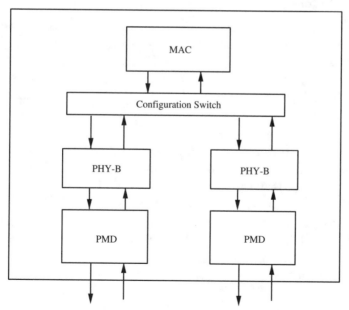

Figure 3.8   Single MAC dual attachment station.

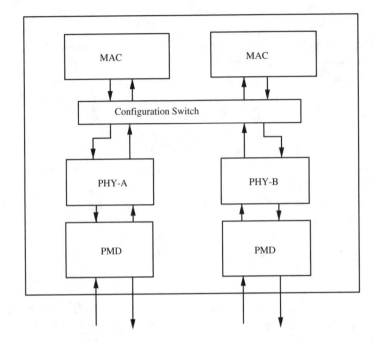

Figure 3.9   Dual MAC dual attachment station.

3.3.3 Concentrators

A concentrator is a multiport device much like a 10BaseT Hub or a Token-Ring media access unit (MAU). A concentrator is very useful because most of the existing wiring in buildings is a physical star topology with one or more central patch-panels and each patch-panel supporting dozens of desktop connections (100 m). In order to form a logical ring without running any additional wire, it is best to connect the concentrator to the patch-panel via short interconnects.

An FDDI concentrator can take on various configurations. It can multiplex a single ring to several desktops (SAC) or it can multiplex two rings (trunk) to the desktop (DAC) as shown in figure 3.10. In low-cost versions, it is even feasible for the concentrator to be the root of the tree (NAC).

Typically a DAC will have optical PMD (MIC, ST or SC connectors) for the A and B ports. It may or may not have fiber for the M ports. Most DACs will initially have mixed media: Fiber A, B ports and Copper M ports. A DAC is required to have a MAC[1].

1. This issue has been fairly controversial in the committee and as a result of compromises, nowhere in the SMT document does it clearly indicate that DACs must have MAC in order to be operational. This is embedded in the CFM state machines.

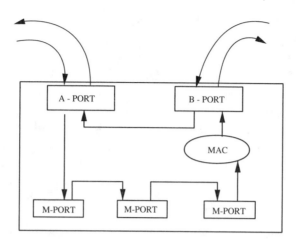

Figure 3.10   Single MAC dual attachment concentrator.

A low-end DAC will support from eight to 20 ports in a standard PC chassis with the M ports configurable on one path (primary) only. It will have one MAC available on the primary ring. Most DACs will provide SNMP support. As the chipsets integrate further and board-space considerations lessen, the number of concentrator ports supported in a chassis will increase.

At the higher-end of the spectrum a DAC will support from 8 to 50 ports configurable on primary, secondary or local rings. Multiple MACs may be available for graceful insertion or deinsertion of nodes. The M ports may be bank-switchable to optimize bandwidth utilization. Ring monitoring and test-ring features may be provided. A terminal might display a ring map reflecting the state of the ring.

Another interesting configuration for the high-end concentrators would be to package them as intelligent hubs in what is referred to as a *collapsed network backplane*. Thus, FDDI would share a high-speed backplane with Ethernet or Token-Ring and allow for intelligent switching of packets. This allows for excellent domain isolation.

To provide further fault-isolation and extend network size without increasing the number of stations on the trunk ring, SACs can be used in a concentrator tree. This provides easy domain structuring and the ability to create a structured hierarchy of FDDI nodes.

SACs are generally copper-based and provide a single ring connection to the desktop. SACs have an S port and multiple M ports. SACs may or may not have MACs. Typically, it is desirable to have a MAC in a SAC if it is to be used in a concentrator tree. Again, depending on cost-performance considerations, SACs may have additional features such as SNMP support, statistical reporting and multiple paths (multiple domains).

3.3.4 Optical Bypass Switches

Optical bypass switches or relays provide a mechanism to maintain ring continuity in the presence of a fault or when the transceivers at a station are powered off. The optical bypass switch allows the light to bypass the optical receiver in a node.

The main advantage of using an optical bypass switch is that the dual-ring topology can be maintained in the presence of faults at individual stations or when stations on the dual ring are powered off. This is important for fault-tolerance. When an actual cable-break occurs, the dual ring can wrap around the fault without isolating any station. Without the optical bypass, multiple cable breaks could cause the logical ring to fragment into different networks.

There are several disadvantages of optical bypass switches which should be considered when using bypass relays:

• Bypass relays cause a power penalty by introducing additional loss in a link. Bypass relays are not amplifying devices. The network operation is compromised if multiple stations (greater than four) are bypassed such that the maximum distance between two active stations exceeds the transceiver specifications.

• Bypasses, like any other mechanical device reduce the reliability of the network.

• Bypasses cannot be used in single attachment stations or connections. This can cause the CMT to fail.

3.4 NETWORK TOPOLOGIES

Although the simple rule of forming an FDDI network requires that there be no more than two data paths–primary and secondary, the rule does not preclude other private paths from being supported in a concentrator. Additional private paths called local paths may be implemented. However, local paths are beyond the scope of the standard and as such cannot be managed by the standard SMT. Many vendors implement concentrators with multiple paths. In such vendor equipment, the local path is used for management, debugging and *graceful insertion*. These functions are discussed in detail in later chapters.

The dual ring specification provides for reconfiguration of the network in the presence of a single fault (figure 3.2). If a fault occurs, this reconfiguration is automatically performed by SMT.

The following subsets of the dual ring with trees are legal topologies:

• Dual rings without trees

• Wrapped ring with trees

- Wrapped ring without trees
- Single tree

3.4.1   Dual Ring with Trees

The dual ring with trees affords the most fault-tolerance and fault-isolation (figure 3.1). It can also offer higher bandwidth (200 Mbps) as each of the two rings can be used for data transmission. This architecture serves well in a large environment. In a smaller network, this topology can be expensive to install (because of the fiber costs in the trunk) and manage (complexity of the topology).

3.4.2   Dual Ring without Trees

The dual ring without trees implies that all end stations are on the dual ring. Hence, all end stations have the more expensive dual-attachment connections. This would be a typical environment for a mainframe connectivity, backbone of servers, or critical applications (figure 3.11).

The wrapped ring with trees is a degenerate case of the dual ring with trees (figure 3.1). This topology should only exist temporarily because of a fault in the network. It is not cost effective to have a wrapped ring with trees as it calls for dual attached stations to act as single attached stations. This topology is more efficiently constructed by a single tree architecture (figure 3.12).

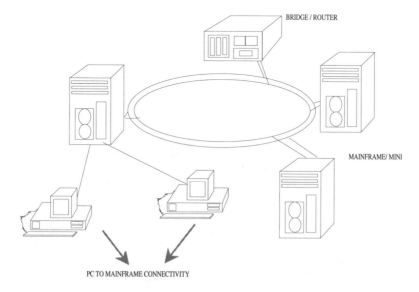

Figure 3.11   Dual ring of dual attachment stations used as a backbone.

3.4.3   Wrapped Ring without Trees

The wrapped ring without trees is a degenerate case of the dual ring without trees (figure 3.6). This topology also should only exist temporarily because of a fault in the network. It is not cost effective to have a wrapped ring without trees as it calls for dual attachment stations to act as single attached stations. This topology is more efficiently constructed by a single tree architecture (figure 3.12).

3.4.4   Single Ring of Trees

This topology is logically equivalent to a single tree architecture. Depending on the environment, it may be more practical to have a single ring of trees or a single tree. In the first case, the geographical spread may be large enough to require multiple concentrators connected in a ring. A single tree, with cascading concentrators is a lower-cost yet effective solution where the location of the patch-panels, or the need to cascade domains is critical.

3.4.5   Fault-Tolerance: Dual-Homing

Several FDDI vendors recommend using only concentrators in the dual ring. This allows additions to the FDDI network very easily as new stations can simply be plugged to the concentrator M ports. Attaching end-stations directly to the trunk ring as DAS stations reduces ring stability and impacts the security of the network. On the other hand, attaching all stations as SAS connections for important network devices such as servers, routers and

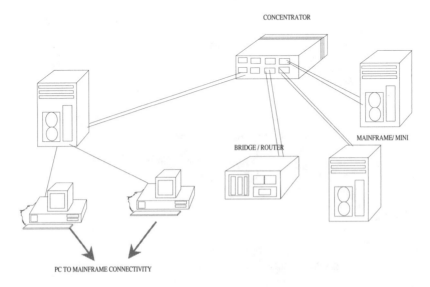

Figure 3.12   Single ring of single attachment stations.

bridges is also risky. In the event of a single cable fault on the SAS link, the server, router or bridge is isolated and can cause severe problems.

A compromise solution is to connect important network devices as physical DAS and logical SAS. The two links are connected to M ports in different concentrators, preferably using separate cable bundles. At any given time only one of the links is active. The other link is maintained in a *withheld* state. A connection is withheld if after initialization and testing, the port on the link is maintained in a standby mode. If the other link fails, then this link is activated and the connection to the network is maintained.

Some workstation vendors provide two SAS adapters rather than a DAS adapter. One of the connections is a backup and can be connected via a different cable to a different concentrator. In the case of a link or concentrator failure, the other SAS connection becomes active. This configuration is different from a DAS configuration because the SMT is simpler and the port type is still S. However, this connection is similar to a DAS with A to M and B to M connections where one of the connections is withheld. Only one connection is active at any time and the other connection becomes active whenever the first one becomes inactive (due to some fault).

3.4.6   Workgroup Environment

FDDI has definitely found a niche in the high-end backbone applications. However, in order for it to move to the desktop three things need to happen:
low cost
lower cost
still lower cost
This may require a very close look at the price/performance requirement of the desktop. Some of the features which add to the expense are:
• Currently optical transceivers constitute 30 to 50% of a SAS connection cost and about 40 to 65% of the concentrator connection. Moreover, the cost of installing fiber is prohibitive.
• Current FDDI chip-set pricing constitutes 25 to 40% of a SAS connection cost and about 25 to 45% of the concentrator connection.
• A MAC in a concentrator constitutes 10 to 15% of the hardware cost and the software for the MAC (SMT protocols, SNMP etc.) adds another 10 to 15% to the cost.

Copper PMD addresses the transceiver cost issue (although the savings in using installed wiring itself is substantial). FDDI chip-set prices are being relentlessly driven down to sub-hundred dollar SAS connections. These combined efforts are making it more attractive to have FDDI on the desktop.

Although a concentrator tree topology is a cheaper solution than a complete DAS network, a DAC or a SAC topology is not necessarily the cheapest.

Typically, most networks in a CAD/CAE environment are anywhere from 4 to 20 nodes and operate in clusters. These networks require the bandwidth without sophisticated management capabilities. This can be ideally provided by a null attachment concentrator (NAC). A NAC is a multiport device without any MACs or A, B or S ports. It consists solely of M ports and optionally a MAC for remote management. A NAC is similar to a low-cost Token-Ring MAU or a 10BaseT hub–an active star-topology distribution point. It makes use of existing wiring (because of small distances and copper interface) and provides 100 Mbps shared bandwidth to the attaching SAS connections.

A NAC is thus simple, robust and cheap. Typically, a NAC can be daisy-chained to extend the network. A NAC based network may contain routers, bridges and gateways. Most vendors provide an upgrade path from a NAC to a NAC with a MAC, to a SAC or even a DAC. As FDDI chipset prices decrease the price difference between a NAC and a SAC may become negligible but currently there is a substantial difference in the price of a NAC and a SAC.

Figure 3.11: An example of a topology where both A and B ports are active in tree mode leading to multiple (3) paths being created when only two were desired

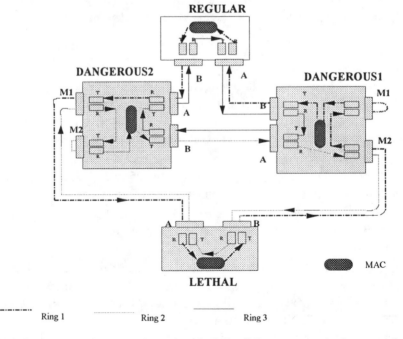

Figure 3.13   An example of a topology where both A and B ports are active in tree mode leading to multiple (three) paths being created when only two were desired.

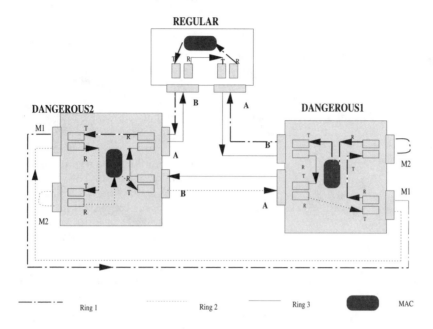

Figure 3.14   An example of multiple rings (three) created when M to M connections are formed.

Table 3.1   **Connection Criteria and Rules**

Connection	Validity	Comments
A to A	Legal but undesirable	This creates a twisted ring as described in figure 3.5.
A to B	Legal	This is the normal trunk ring DAS (or DAC) connection.
A to M	Legal if B to M not simultaneously active	This connection can create multiple rings if B to M also active. However, it can be used for redundancy and fault-tolerance purposes. This is a DAS connected as a SAS in the tree.
A to S	Legal but undesirable	This connection is undesirable because it causes a wrap and reduces two rings to one.
B to A	Legal	This is the normal trunk ring DAS (or DAC) connection.
B to B	Legal but undesirable	This creates a twisted ring as described in figure 3.5.

Table 3.1 **Connection Criteria and Rules (Continued)**

Connection	Validity	Comments
B to M	Legal if A to M not simultaneously active	This connection can create multiple rings if A to M also active. However, it can be used for redundancy and fault-tolerance purposes. This is a DAS connected as a SAS in the tree.
B to S	Legal but undesirable	This connection is undesirable because it causes a wrap and reduces two rings to one.
M to A	Legal	This is not the same as an A to M connection because the M port is in a concentrator whereas the A port is in a DAS. Some concentrators may choose to disallow M to A connections altogether. Other concentrators which allow M to A connections assume that the DAS (or DAC) is ensuring that both A and B are not active in *tree* mode (figure 3.13).
M to B	Legal	This is not the same as a B to M connection because the M port is in a concentrator whereas the B port is in a DAS. Some concentrators may choose to disallow M to B connections altogether. Other concentrators which allow M to B connections assume that the DAS (or DAC) is ensuring that both A and B are not active in tree mode.
M to S	Legal	This is the normal tree connection.
M to M	Illegal	This is the only illegal topology which creates a tree of rings architecture, leading to multiple (fewer than two) paths (figure 3.14).
S to A	Legal but undesirable	This connection is undesirable because it causes a wrap and reduces two rings to one.
S to B	Legal but undesirable	This connection is undesirable because it causes a wrap and reduces two rings to one.
S to M	Legal	This is the normal tree connection.
S to S	Legal	This connection also forms a single ring with two SAS or two SAC (and many SAS) or a SAS and a SAC (and many SAS).

FDDI Media Access Control

4.1 FDDI MEDIA ACCESS CONTROL (MAC)

In this chapter, the MAC sublayer of the data link layer is discussed in detail. The other sublayer of the data link layer is the logical link control (LLC) which is common to all local area networks. The MAC layer interacts with the LLC, PHY, and SMT layers and performs important functions such as framing, addressing, error checking, token generation, transmission time control, and fault isolation.

4.2 MAC FUNCTIONS

The IEEE 802.1A specifies the OSI reference model for a LAN. A LAN implements layer 1 (physical) and layer 2 (data link) of the seven layers. The data link layer for every LAN consists of: logical link control (LLC) and media access control (MAC) as shown in figure 4.1.

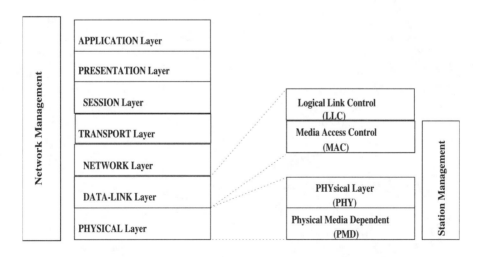

Open Systems Interconnect Reference Model   FDDI specifications mapped to OSI

Figure 4.1 FDDI and the OSI reference model.

The MAC typically provides at least the following four functions:

• Ordered and controlled access to the media by the nodes on the network

• Address recognition

• Fair access to the media: No node should utilize an unreasonable proportion of the bandwidth at the expense of other nodes.

• A bound on the network access delay for a node

• Error checking mechanism

All LANs provide the above functions. Some LAN MACs provide enhanced functionality. For example, a CSMA/CD network such as IEEE 802.3 does not provide a guaranteed bound on the network access delay for a packet whereas the Token-Ring and FDDI MACs provide a bounded access time class of service. The FDDI MAC provides the above four basic functions and the following additional functions and features:

• Multiple classes of traffic

• Dual mode of operation

• Low-level acknowledgment scheme

• Multiple address recognition scheme

• Error detection mechanisms such as frame check sequence (FCS) and error indicators

• Frame handling

• Interface to station management and other upper-layer protocols

The LLC sublayer is similar for most LANs. Each LAN uniquely defines a MAC layer protocol. Thus, the Token-Ring MAC (IEEE 802.5) is different from the Token-Bus (IEEE 802.4) and CSMA/CD (IEEE 802.3) MAC although they share a common LLC. Similarly, the FDDI MAC is unique in its functions and features, yet it shares a common interface to the LLC sublayer (figure 4.1). According to the IEEE 802.1A (network architecture) specification, a network consists of a protocol stack on top of different *subnetworks*, where subnetworks are the different media access control and physical layer such as Token-Ring, Ethernet (or IEEE 802.3), FDDI, and so on.

4.3   HISTORY OF THE FDDI MAC PROTOCOL DEVELOPMENT

The protocol adopted for the FDDI network access mechanism is the Timed Token Rotation protocol. This protocol was initially proposed by Robert Grow [1] in 1982 for a ring architecture. At the time, Grow was with Burroughs corporation. Around the same time, Sperry had started development on a high speed network. The initial work at Sperry was done by James Torgerson and James Hamstra. On completion of the initial work, Sperry decided to take their proposal to ANSI for development of a high speed network standard. Advanced Micro Devices was the silicon partner in the Sperry proposal. Thereafter, Sperry and Burroughs merged and several other companies also became involved in the development of a 100 Mbps standard. Today, the interest in FDDI is world-wide and a wide range of products is available in the market.

The FDDI MAC was developed by the ANSI X3T9.5 committee. The work on the MAC began in 1984. It became an ANSI standard in 1987 (X3.139-1987) and an ISO standard in 1989 (ISO 9314-2). The implementation of FDDI networks was greatly facilitated by the early silicon solutions. AMD was the first such vendor to implement an FDDI chipset concurrent to the development of the MAC and PHY standards. Several implementations have since been introduced in the market that comply with the ISO 9314-2 standard.

The FDDI MAC standard (ISO 9314-2) does not specify the behavior of a station MAC in the presence of bridges. At the time, it was decided to leave it to the IEEE 802.1D bridging committee to clarify the bridging requirements. However, several key issues remained unspecified. Specifically, the handling of the frame status indicators and the stripping of forwarded frames. Moreover, minor bugs and loopholes were found in ISO 9314-2 after several implementations were deployed in the field. In order to fix the problems and to facilitate the use of the FDDI MAC with the Hybrid Multiplexer of FDDI-II as a Packet MAC, an additional MAC specification called MAC-2 was developed. The MAC-2 is an enhancement of the ISO 9314-2 MAC specification with the following additions and / or fixes:

- Specifies the behavior of the FDDI MAC in the presence of bridges

- Provides an interface to the Hybrid Multiplexer of FDDI-2

- Fixes minor bugs and loopholes in the ISO 9314-1 MAC standard

The draft proposed ANSI (dpANS) MAC-2 specification is expected to become an ANSI standard by the end of 1992 and an ISO standard by 1993.

4.4 NETWORK ACCESS MECHANISM

The protocol adopted for the network access mechanism is the Timed Token Rotation (TTR) protocol. Conventional circuit-switched networks are designed for low latency, low data rate applications, and are generally not suitable for handling *bursty*[1] data traffic. Local area networks are optimized for bursty data traffic and are not well adapted to handling digital voice. In the initial feasibility study done at Sperry and Burroughs, it was felt that a

---

1. Traffic is considered bursty when it follows a pattern of light traffic for long periods followed by heavy traffic for short periods.

ring architecture would be more amenable to providing a low access delay, high throughput, and fault-tolerant channel. At the time Grow submitted his proposal, a token-based ring architecture was seen as more appropriate than a slotted ring architecture.

There were three classes of service in the proposed Timed Token protocol:

- Class 1: Guaranteed Bandwidth, *Synchronous*

- Class 2: Shared Bandwidth, *Interactive*

- Class 3: No guarantee of Bandwidth, *Batch*

The first class of service is designed for applications requiring a guarantee of bandwidth and/or deterministic delay and *jitter* characteristics. Jitter is the variable delay introduced by the network in transmitting each packet. Therefore, if the first packet takes x ms after being queued and the second packet takes y ms after being queued then the jitter is x − y ms. The guaranteed bandwidth and deterministic delay and jitter characteristics are useful for synchronous information such as digital voice. Process control applications, where minimal queuing delay is important, can also be supported by this class of service.

The second class of service is designed for traditional data communications where a minimum throughput is provided but absolute guarantee of bandwidth is not. This is achieved by creating a pool of bandwidth which cannot be used by class 1.

The third class of service is essentially the left-over bandwidth at light traffic loads. This implies that traffic may be delayed until network load is light. Batch information may be transmitted using this class.

The allocation of bandwidth across the three classes of services was based on the desired token rotation time. A modified version of this protocol was proposed to the ANSI X3T9 committee by Grow and it was adopted as the media access mechanism for FDDI.

4.5 FDDI MAC PROTOCOL

The basic operation of the FDDI MAC protocol is similar to that of a token ring. Access to the network is controlled by passing a *permission* token around the ring. In order to transmit, a station needs to *capture* a token. A token is captured by removing (or stripping) it off the ring. After capturing a token, the station transmits packet data. A station may not hold a token longer than the period determined by the TTR algorithm. At the end of the token holding time, the token is forwarded to the downstream station. In FDDI, it is not necessary for the transmitting station to hold the token until all its transmitted frames have returned. The token is typically issued immediately on completion of transmission. This mechanism allows for more efficient usage of bandwidth and is known as early token release. In 4 Mbps

Token Ring (IEEE 802.5), the token is released only after all transmitted frames are received by the source.

If the downstream station has data to transmit, it will capture the token or else it will repeat it. All stations except the transmitting station must repeat all frames. A station may copy a frame before repeating it or just repeat it. The sender must strip its frames.

Several questions arise: What is a token? Who generates the first token? What happens if the token is lost? What happens if multiple tokens are created?

A token is a special type of permission frame. In Token-Bus, a token contains a source and destination address field. In Token-Ring, while the token does not contain an address field, it does contain priority and reservation fields. In FDDI, a token is a very small frame. It contains a frame class field (an octet) which distinguishes this frame as a token. There is no other field besides a starting and ending delimiter. The symbols denoting the token are JK80TT. The J and K symbols represent the start delimiter and T represents the ending delimiter. Hex 80 is the value of the frame control fields. Symbols are explained later in chapter five. The frame control field is explained later in this chapter.

The first token is typically generated by a designated station. In the literature of token-based networks, various schemes have been suggested and implemented for generating the token and controlling the access, ranging from a central monitor (as in Token-Ring), to a bidding scheme. FDDI implements a bidding scheme. In this scheme, every station bids for how fast it wants the token to rotate around the ring. The station bidding for the fastest token rotation wins. If two or more stations have the same bid, the winner is the station with the highest 48-bit IEEE assigned address. Thereupon, the winner transmits the first token which is used by all nodes on the network. The process by which the bidding is accomplished is called the *claim process*. The claim process is explained later in this chapter.

4.5.1 Classes of Service

The FDDI MAC protocol provides multiple classes of service and two mechanisms of operation:

- non-restricted mode
- restricted mode

The non-restricted mode of operation is the normal mode. In order to provide a low access delay service, the TTR protocol specifies two classes of service:

- Synchronous
- Asynchronous

The synchronous class of service has a guaranteed bandwidth and access delay. It is also the highest priority traffic. This is very useful for applications requiring:

• deterministic access to the network

• minimum jitter (variations in inter-frame arrival times at the destination)

• bandwidth guarantee requirements

Typically packet voice, video, and process control applications are well suited for the synchronous bandwidth.

The asynchronous class of service on the other hand, is the remaining bandwidth which is dynamically shared between all the nodes. Thus, asynchronous traffic is sent on the network only if the ring is lightly loaded and is not a guaranteed bandwidth at heavy network loads. Within the asynchronous classes of service FDDI, up to eight levels of priority can be optionally defined. Typically, most data transmissions use the asynchronous class of service.

4.5.1.1 Token Types. Within the asynchronous class of service, two types of tokens can be used:

• non-restricted

• restricted

The non-restricted token is the normal token generated by the claim process and used by all stations. Selected nodes can use the restricted token for extended communication duration's. An example application would be a network with multiple file-servers. Periodically (once a day), the file-servers would need to back-up the data. This can be efficiently executed if the entire network bandwidth was used by the file-servers for an *extended* duration. After the back-up is completed, the token is converted from restricted to non-restricted and normal transmission resumes. During the restricted conversation, the other stations on the network effectively disable the TTR protocol and wait for a non-restricted token to become available. The restricted token class is optional in the standard and rarely used.

4.5.1.2 Basic Ring Operation. In the TTR protocol, station timers are defined, which jointly attempt to maintain a target token rotation time by using the offered network load to regulate the transmission time of each station. Offered load is the actual message load or traffic demand presented to the network. Thus, if a token rotates slowly in one rotation (offered load is high), then the token rotation is speeded up in the following rotation by reducing the transmission time of the offending stations (which offered heavy load on the previous rotation). This speeding up and slowing down of the token around the ring ensures an average token rotation time equal to the Target token rotation time (TTRT) and an upper-bound of 2* TTRT on the access delay for synchronous traffic. The upper-bound on the access delay for asynchronous traffic is much higher and can be as much as $(n - 1) * TTRT + 2*$ ring latency, where n is the number of stations and ring latency is the time it takes a frame to travel around the ring. The proofs for these bounds on TTRT and access delay can be found in [2] and [3].

In brief, the FDDI MAC protocol has a token which is circulated around the ring. Any station wishing to transmit has to capture the token. The station completes transmission when it has no more packets to transmit or its transmission timer expires upon which it immediately releases the token, which can then be captured by the next station. The protocol bounds the transmission time and also provides a hierarchy of services. The synchronous class of service is guaranteed to receive a token at least once in two rotations.

4.6   FDDI FRAME FORMATS AND ADDRESSING

An FDDI frame can be a maximum of 4,500 bytes, which includes eight bytes of preamble. Preamble is idle-fill and is used to delineate back-to-back frames and for clock recovery purposes. The actual information content of the FDDI frame is 4,486 bytes (if using 16-bit addressing) or 4,478 bytes (if using 48-bit addressing). Figure 4.2 shows an FDDI frame.

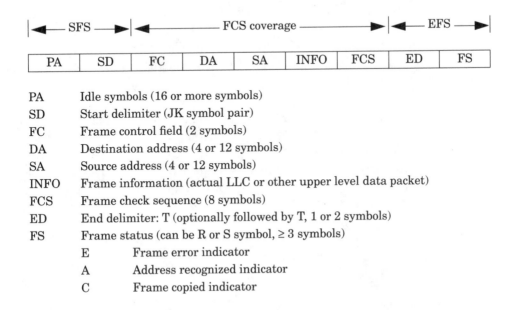

| PA | SD | FC | DA | SA | INFO | FCS | ED | FS |

PA	Idle symbols (16 or more symbols)
SD	Start delimiter (JK symbol pair)
FC	Frame control field (2 symbols)
DA	Destination address (4 or 12 symbols)
SA	Source address (4 or 12 symbols)
INFO	Frame information (actual LLC or other upper level data packet)
FCS	Frame check sequence (8 symbols)
ED	End delimiter: T (optionally followed by T, 1 or 2 symbols)
FS	Frame status (can be R or S symbol, ≥ 3 symbols)

	E	Frame error indicator
	A	Address recognized indicator
	C	Frame copied indicator

Figure 4.2   FDDI frame formatting.

A start delimiter (SD) consisting of two symbols J and K, is used to demarcate the frame beginning. These symbols are specially chosen to be easily recognizable on any bit boundary. Moreover, the probability of random noise corrupting other symbols to J and K is extremely low.

The SD is followed by the frame control (FC) field. The FC field is similar to the type field in Ethernet. FDDI defines several frame classes including LLC (more correctly upper-layer protocol), beacon and claim (MAC frame class), and SMT (subnet management frames). An *implementer* frame class is also defined which can be used by vendors for proprietary protocols.

The FC field is followed by the destination address (DA) and source address (SA) field. FDDI allows two types of addresses to be used: 16-bit and 48-bit. The 48-bit addresses are usually the IEEE assigned addresses. The 16-bit addresses are locally administered and need not be unique across subnets.

FDDI SMT requires the use of long addresses and hence each MAC is capable of at least supporting the long addresses. FDDI frames may be broadcast, multicast, or individually addressed. A broadcast destination address FDDI frame is received by all stations on the ring. A multicast destination address frame is received by all stations subscribing to the multicast or group address. A broadcast address is a degenerate case of a multicast address. A frame with an individual destination address is received by one station only. Except in the case of a *source-routed* frame, the source address always contains an individual address. This is indicated by the individual bit of the 48-bit source address being reset to zero (figure 4.3). In a source-routed frame, the Individual bit of the source address is set to one indicating that the frame is to be source-routed (as opposed to transparent bridging). Source routing is further discussed in chapter 9.

The address fields are followed by the information field. This may be an LLC frame encapsulating an upper layer protocol. A popular upper-layer protocol is the Internet protocol (IP) used in conjunction with the transport protocol Transmission Control protocol (TCP) or the User Datagram protocol (UDP). IP is a network-level protocol and its protocol data unit (PDU) is called an IP fragment. RFC 1151 (or the latest version), "IP over FDDI" specifies that

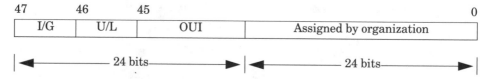

I = 0; G = 1; Individual or Group address;
U = 0; L = 1; Universal (IEEE assigned) or Local address

Figure 4.3 48-bit addresses.

the maximum size of the FDDI information field shall be 4k bytes. An additional 256 bytes of overhead (MAC + LLC + IP + TCP) are provided for leading to a total frame length of 4352 bytes. Within the transport PDU are encapsulated file transfer, email and other application-specific messages.

The information field is followed by a frame check sequence (FCS) which is automatically generated by the MAC on every frame. The FCS field is four bytes wide. The FCS checks for errors in the frame up to and including the FCS field. Every incoming frame passes through the FCS logic which continuously computes the FCS using a standard 32-bit cyclic redundancy check (CRC) polynomial. The FCS computation is stopped when the end of frame indicator (T symbol) is reached. The calculated FCS is compared to the last four octets before the T symbol. If the two do not match, an error has been detected. The FCS algorithm can detect most errors. More information about the FCS algorithm can be found in ANSI X3.139 (FDDI MAC) or ANSI X3.66 (ADCCP).

FDDI does not require the information field to be word (32-bit) aligned. It is up to the user to transmit word-aligned data or odd number of data bytes. Most upper-layer protocols will word-align the data by padding at the beginning or the end of the information field.

Following the ending delimiter T, are the frame status indicators (as shown in figure 4.4). They indicate the current status of the frame: whether an error was detected by the FCS (E indicator), whether the destination station recognized the frame (A indicator), and if the frame was copied by the

Frame Status	E	A	C
Just after transmission	R	R	R
After leaving intermediate station without errors	R	R	R
After leaving intermediate station finding an error	S	R	R
After leaving destination Finding no error	R R	S R	S (if copied) R (if not copied)
After leaving destination With error	S S	R S	S (if copied) R (if not

Figure 4.4   Frame status indicators.

destination (C indicator). These three indicators (or symbols) E, A and C can either be Set (S) or Reset (R). R and S are two defined symbols. Optionally, additional indicators may be defined which may be used to operate a proprietary protocol. The setting of the E, A and C indicators is as shown in figure 4.4.

4.7   NETWORK TIMERS, VARIABLES AND FLAGS

The TTR protocol uses a number of variables and timers at each station to determine the length of time the token may be held for transmitting frames of a given class. Some of the more important variables (and or flags) and timers are:

4.7.1   Variable: Token Rotation Time Requested (T_Req)

T_Req: Token Rotation Time **Req**uested by each station. During the claim process, each station puts in a request for how fast it wants the token to rotate. In other words a station desiring to see a token at least once in x ms will put in a T_Req = x ms. The station requesting the fastest token rotation time is the winner as it satisfies every other station's request. For example, in a three-station ring, if station A bids 10 ms, station B bids 20 ms, and station C bids 30 ms, the winner is station A. Stations B and C cannot win because they fail to satisfy station A. The T_Req value is decided mainly by applications which require a time-bound on the network access delay. For asynchronous applications, the selected rotation time is immaterial.

4.7.2   Variable: Token Operational Time (T_Opr)

T_Opr: The **Oper**ational **T**ime value agreed to by all stations. The T_Opr is the value of the bid (T_Req) put in by the winning station. This winning bid becomes the target token rotation time (TTRT) and is loaded into the T_Opr register. Thus, this value is the highest numerical value of all the requested values; highest because the value is actually implemented in two's complement logic in the MAC standard, and in two's complement logic the smallest numerical value is the largest number.

4.7.3   Variable: Token Rotation Time-Maximum (T_Max)

T_Max: Maximum value of TTRT supported by a MAC. Why should there be a limit on the value of TTRT? The answer lies in the desire to have a upper-bound on the time it takes for a ring to recover. The ANSI committee decided that T_Max should be several times the maximum ring latency.

Thus, T_Max is a very large number (> 100 ms) and is sufficiently padded that the ring initialization will be completed within T_Max. If the ring initialization does not complete within T_Max, there is a problem with the ring.

The default value of T_Max is 165 ms. Many implementations have different granularity and hence are approximations of 165 ms. For instance, some implementations may have a value of 167.772 ms.

4.7.4   Timer: Token Rotation Time (TRT)

TRT: **T**oken **R**otation **T**imer. A timer used to measure the time between successive arrivals of the token at the station. The TRT is loaded with the T_Opr value which is the two's complement of the TTRT. The TRT is reset each time the token arrives earlier than the TTRT. If the TRT expires before it sees the token, the token is considered late. The MAC then increments (or sets) a Late_Ct counter (or flag) and reloads TRT with T_Opr and the TRT starts counting up again. If the token arrives at a station after the TRT has already expired once and the timer has reloaded, the TRT is not reset. This has the effect that the station's TRT will expire a little earlier on the next token rotation. If the token does not arrive within two consecutive expirations of the TRT, the station reinitializes the ring (see examples later in the section).

On reset when the ring is operational, the timer is reloaded with the T_Opr value. On reset when the ring is not operational, the TRT is loaded with T_Max.

4.7.5   Timer: Token Holding Time (THT)

THT: **T**oken **H**olding **T**imer. This timer is used to control the amount of time that a station can transmit asynchronous class frames. It is loaded with the value remaining in the TRT timer every time the token arrives at the station. If the token rotates faster around the ring (ring is lightly loaded; not many stations are transmitting) there will be more time remaining in the TRT when the token returns to this station. Therefore, a larger time value will be loaded into the THT timer allowing this station to transmit more. A station transmits asynchronous frames until the THT timer expires. This timer is only enabled once a station sees a token, enough time remaining in the TRT (TRT not expired), and the token is captured. This timer is for asynchronous transmissions and is enabled at the end of synchronous transmissions as synchronous transmissions are higher priority. If the timer expires in the middle of a transmission, the current transmission is completed before releasing the token.

4.7.6   Flag: Late_Ct

If the TRT counts up to the TTRT and the token has not been received, the token is *late* and the Late_Ct flag is set. When a token is late, only the top-priority traffic, which is synchronous, can be transmitted. No asynchronous traffic is permitted to be transmitted on a late token arrival.

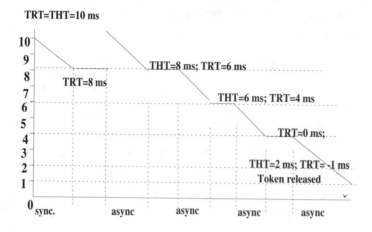

Figure 4.5 A node receives a token when value remaining in TRT = 10 ms. It loads that value into the THT timer which is frozen during synchronous transmissions. At the end of sync. transmissions, it can transmit asynchronous traffic for THT = 10 ms. It divides this up amongst the multiple asynchronous priorities. It transmits the highest priority first and then the remaining priorities depending upon the time left.

4.7.7 Timer Valid Xmission (TVX)

TVX: **T**imer **V**alid **X**mission. This timer is used to place a bound on the time a ring will remain operational without a station receiving valid frames. In other words, "how long does it take for the ring to recover from a lost token?" If a token is lost, no station can transmit. Therefore, there is no valid data on the network. In the absence of a token, the TRT timer in each station will expire twice before ring recover is initiated. In the TTR protocol, if no valid data is received for 2* TTRT then the ring should initialize itself. Twice TTRT in the worst case can be 2*165 (default T_Max value) ms = 330 ms. In order to recover more quickly from a lost token, a TVX timer is defined. This timer times the duration between valid frames. It is reset on receipt of a valid frame. If the TVX timer expires, no valid frame including a token has been received indicating that the token is lost or corrupted and recovery procedures are initiated.

The value of TVX is calculated from the time it would take a maximum-sized frame (4,500 bytes, F_Max = 0.361 ms) to go around a maximum sized ring (twice D_Max = 2 × 1.773 ms). This is a default of 3.4 ms[2].

---

2. The MAC standard incorrectly specifies the TVX to be 2.5 ms.

4.7.8   Variable: Token Rotation Time-Minimum (T_Min)

T_Min: The minimum value of TTRT supported by a MAC. This attribute, T_Min, has been a little contentious in the committee. One camp views this attribute as nothing more than a practical bound on the TRT to ensure that the ring operates in a stable manner. Another camp views this as a constant which should be 80 ns (the byte granularity in FDDI) in order to maintain ring stability under certain boundary conditions.

Practically, it can be thought of as the lowest common denominator of the granularity's of T_Req of different implementations. This has been selected to be a default of 100,000 symbol times or 4 ms.

4.8   OPERATION OF NETWORK TIMERS: TRT AND THT

The two basic timers required for the operation of the FDDI timed token rotation protocol are TRT and THT. The TRT and THT are defined earlier in this chapter. The TRT timer measures the actual rotation time of the token on the ring. It resets to T_Opr every time that a token is received early (i.e. Late_Ct is = 0). If a token is late on a given rotation, the TRT rolls over and accumulates the *lateness* of the token and is not reset to T_Opr. Hence, every time that a token is late, the value in TRT is less than T_Opr. This will cause the TRT to expire earlier on the following token rotation unless the token rotates faster and is received before the TRT expires again. If a token is received when the TRT has expired, no asynchronous transmissions are allowed. However, synchronous transmissions are permitted on every token rotation up to a theoretical maximum of TRT. The TTR protocol guarantees a maximum access delay of 2*TTRT for synchronous data. However, this guarantee is only met if the total synchronous transmissions at each station on any given token rotation are limited to TTRT. The FDDI MAC protocol does not provide any mechanism of synchronous bandwidth allocation across all the stations. This is left up to the user. Therefore, extreme caution should be exercised when using the synchronous bandwidth. Over-allocation of the bandwidth can cause the ring to crash frequently.

If a token is received early (token rotates faster than the target TRT), the value remaining in TRT is loaded into the THT and the THT is used to time the asynchronous transmissions. A station can transmit up to the THT value. The maximum value that can be loaded into the THT is TRT. THT is typically disabled during restricted operation and synchronous transmissions. THT is reset to zero when the token leaves the station.

The operation of the TRT and THT can be best explained by the following example (figures 4.6 to 4.12).

We assume a T_Opr = 10 ms. Assume for the purposes of this example that each station has an infinite stream of data available at the asynchronous

queue for transmission. Also assume at the instant of the snapshot, the ring
has been initialized and the various TRTs are offset by the ring latency
between the nodes. Since we are assuming only asynchronous transmissions,
the other timer of interest is THT. At this time, the value of THT in station A,
B and C is 10 − x, 0 and 0 ms respectively, where x is the ring latency. Assume
station A is the first station to transmit.

• Figure 4.6: A begins to transmit and since it has an infinite queue of
data available for transmission, it transmits until THT has counted down to
zero (a maximum of TRT − x ms).

• Figure 4.7: The TRT has expired and rolled over to a value of 10, 10 −
x/3, and 10 − 2x/3 ms at A, B and C respectively. The Late_Ct is set in all the
stations and station A has just released the token. The THT is zero at all stations. The ring is filled with frames from station A or idles or fragments of
stripped frames (mostly stripped by station A).

• Figure 4.8: When the token reaches station B, it is not able to capture
the token because the token is already late (Late_Ct = 1) and TRT has rolled
over. Station B does not reset the TRT and THT is not loaded with the value of
TRT. The late token resets the Late_Ct to zero and is then released.

• Figures 4.9 and 4.10: Stations C and A cannot capture the token either
and are forced to repeat the token after clearing the Late_Ct. Thus, in the first
token rotation only station A gets to transmit. The other stations repeat the
token as it is already late. In the following token rotation, the token can be
captured by station B which is downstream of station A.

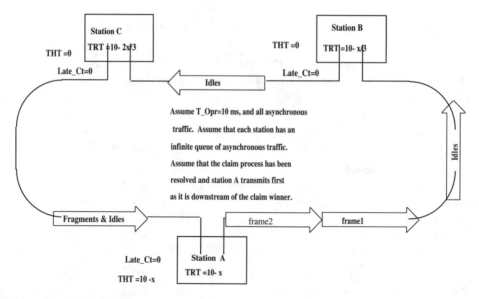

Figure 4.6   TRT and THT.

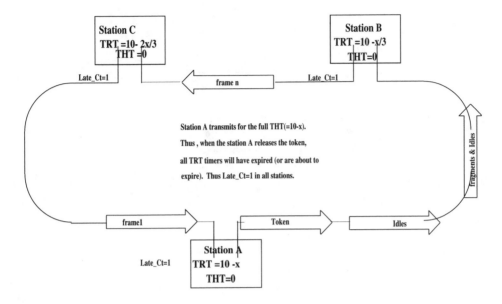

Figure 4.7   TRT and THT.

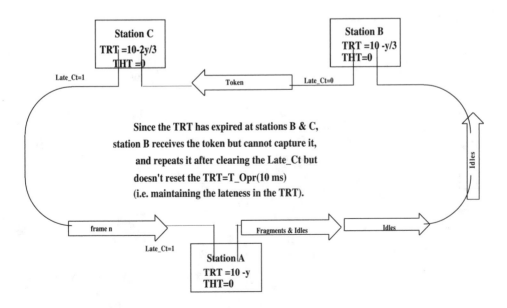

Figure 4.8   TRT and THT.

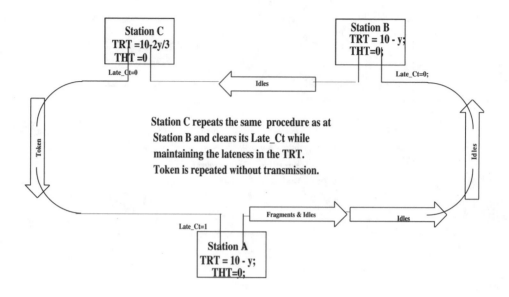

Figure 4.9   TRT and THT.

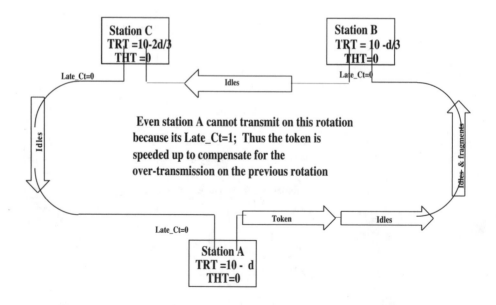

Figure 4.10   TRT and THT on a ring.

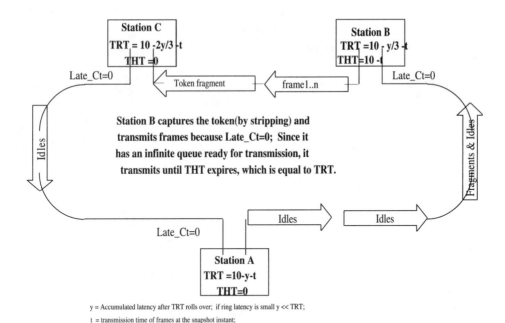

Figure 4.11   TRT and THT.

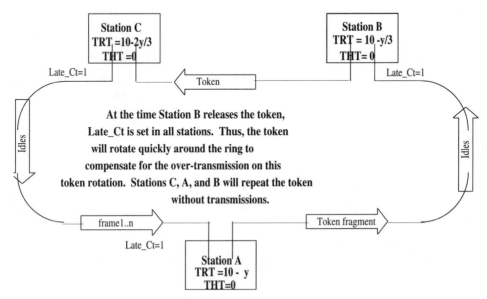

Figure 4.12   TRT and THT.

Thus, station A can transmit on the first rotation, station B can transmit on the second token rotation, and station C can transmit on the third token rotation. This leads to some interesting results for the case of a network with heavy loads.

According to Jain's analysis [Jain90], it is possible that it could take up to $(n - 1)*T + 2*D$ before a station can access the ring; where n = number of stations, T = TTRT, D = ring latency.

Thus, in a 101 station ring with a TTRT = T = 165 ms, and a ring latency of D = 1 ms, it can take up to 16.5 seconds before each station could transmit! Therefore, even though fairness is maintained under heavy loads (every station gets to transmit), the time period within which each station gets to transmit is extended by orders of magnitude which might make it unacceptable for certain applications. Jain maintains that a TTRT of 8 ms serves the purpose well and recommends it as a default. This would mean a station would still have to wait up to 0.79 sec before transmitting in the worst case. However, this does not take into account the synchronous transmissions and the possibility of synchronous and asynchronous transmissions. TTRT should be set based on the latency requirements, type of traffic, ring latency (and number of nodes) and mix of traffic.

4.9   OPERATION OF NETWORK TIMERS: TVX

It is possible for a token to be lost. Although this is a low-probability event, it can still occur. If a token is lost, in the absence of other mechanisms,

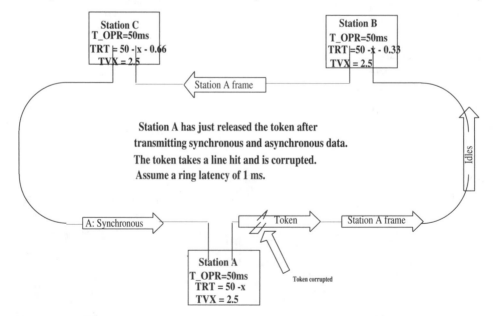

Figure 4.13   TVX.

stations can take as much as 2*TTRT, which is 330 ms in the default case before reacting to it. Three-hundred-thirty milliseconds is a long period of time to be wasted by a 100 Mbps network. To overcome this situation, a valid transmission timer has been defined (TVX) which times the interval between consecutive valid frames. A valid frame must have an FC, DA, SA, optionally information field, correct FCS and FS E-indicator reset (please refer to the ANSI MAC or MAC-2 document for a stricter definition of a valid frame).

If the token is lost or corrupted, no station is able to transmit and gradually the ring becomes devoid of valid frames. All valid frames are eventually stripped by the sending stations and the ring will degenerate to consist of idles and fragments. The network is operational without any data transmissions. The TVX timer is implemented in order to recover from a lost token more quickly than the 2*TTRT (see sec. 4.7.7) provided by the TTR protocol. If any stations TVX timer times out, that station initializes the ring.

The working of the TVX timer is illustrated by figures 4.13 to 4.16.

4.10 RING INITIALIZATION AND CLAIM PROCESS

FDDI is a self-constructing and self-healing (to an extent) network. This means that if a network is designed to have 200 nodes, it is not required to have all 200 nodes active for the network to be operational.

Typically, on power-up, every node tries to form a connection with its upstream and downstream neighbor. If the neighbors are also attempting to

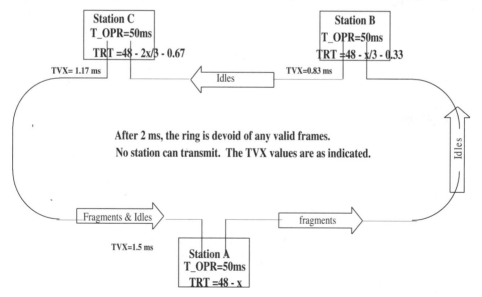

Figure 4.14 TVX.

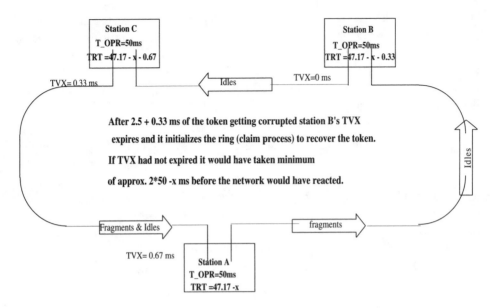

Figure 4.15   TVX.

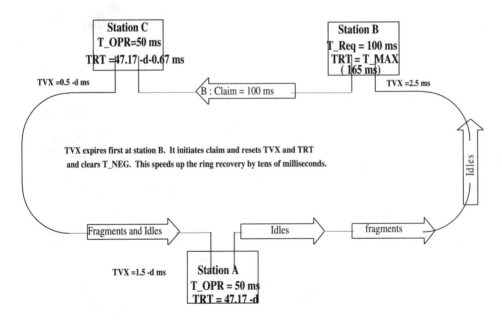

Figure 4.16   TVX.

form a connection, a connection is established between the two neighbors. The chain keeps on evolving until there are no more neighbors desiring to go active or all nodes in a network are active, at which time the network is completed.

The process by which each node forms a connection is a part of SMT called Connection ManagemenT (CMT). This process involves the PHYsical layer interactions. After neighboring PHYsical layers have initiated the dialog and established a connection, the MAC is available for placement on the *logical ring*. The CMT portion of connection establishment exchanges preliminary information about the ports (such as port type, link quality, etc.). Once the CMT becomes active and the connection is established, the MAC is available to be placed on the logical path and be part of the logical ring.

The process of putting the MAC on the logical path and making a logical ring available for the transmission of PDUs is called the *claim* process.

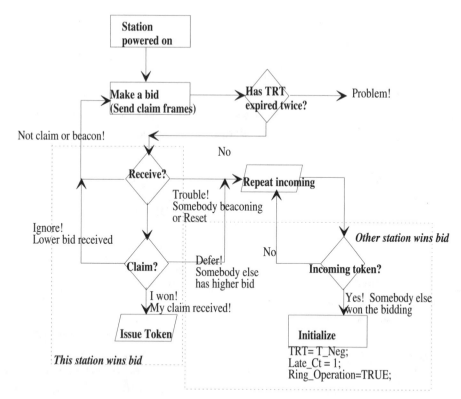

Higher bid implies twos complement higher; i.e. 20 ms in twos complement is higher than 40 ms

Figure 4.17a   Flow-chart for claim process.

1	1	2, 6	2, 6	4	2			bytes
SD	FC	DA	SA	TREQ (TBID)	T	E	A	C

FC := Frame Control :: 1L00 r011; 0xC3 (if Long address used) else 0x83 (if short address used); r is xmitted as 0;

DA:= Destination Address :: DA = SA always;

SA:= Source Address :: A station must recognize a short address even if it uses a long address;

TREQ:= The requested target token rotation time (TTRT) in unsigned two's complement;

E:= Error Indicator :: This indicator must be received reset for the claim to be valid;

A:= Address Recognized Indicator :: This indicator should not be set by any station;

C:= Copied Indicator :: This indicator should not be set by any station;

Figure 4.17b　Claim frame contents.

4.10.1　Claim Process

The claim process is conducted through the use of claim frames (figures 4.17a and b). Claim frames are MAC frames and are restricted to the same FDDI ring. They cannot be bridged or routed as the claim process is local to a ring. The claim process can be divided into two stages: bidding and initialization.

4.10.1.1　Stage 1 – Bidding.　Bidding is a process by which all stations agree on a value for the target token rotation time.

The following sequence of figures 4.18 to 4.21 illustrate the claim process for a three station ring.

In this case, it is the first station to join the network. It should be noted that during the claim process, the TRT timer is reset to the T_Max value (default 165 ms). Each station has a T_Req value which it bids during the claim process. This is called the T_Bid value.

• Station A powers up first and is the first station ready after the initialization of the hardware to join the network. In the mean time, station B and station C have become active in the initial MAC states T0 (transmit idles) and R0 (repeat incoming data) state waiting for network data to trigger the state machines. Station B receives the claim frame from station A which triggers the following:

1.　Station B checks the frame control field and discovers that the FC field equals 1L00 0011; L = 1 = > Long address which indicates that it is a claim frame with long address.

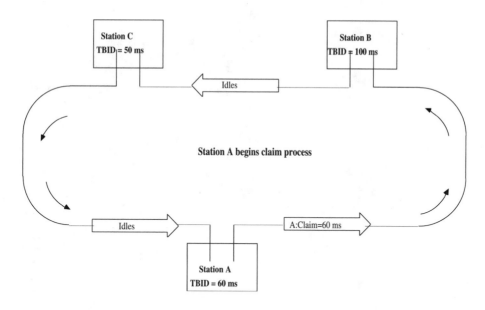

Figure 4.18   Claim process for three stations.

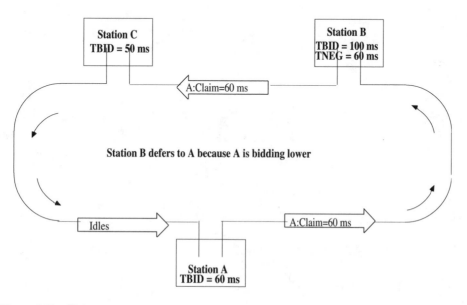

Figure 4.19   Claim process.

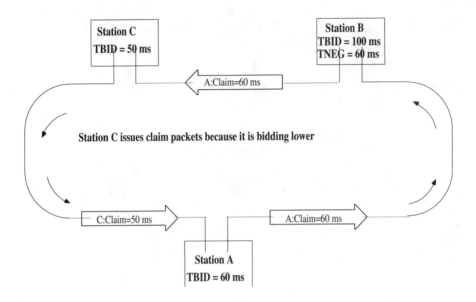

Figure 4.20   Claim process.

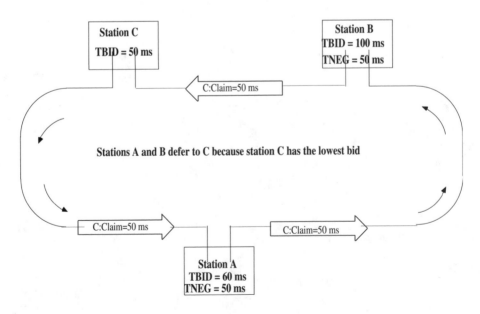

Figure 4.21   Claim process.

2. Then it compares the DA field of the incoming frame to its own address. The DA field of the frame does not match station B's address.

3. Next it compares the SA field to of the incoming frame to its own address (the appropriate long or short address) and sets its H_Flag if the address of the claim frame is higher than the address of this station; otherwise it sets the L_Flag. H_Flag indicates to the MAC transmit and receive state machines that a claim frame with a higher address was received. Similarly L_Flag indicates a lower address.

4. The first four octets of the received frame contain the T_Bid value of station A. It compares the value (60 ms) with its own T_Req (100 ms). Its own value is higher (two's complement lower) and it sets its H_Flag (indicating reception of higher claim).

5. It validates that the frame is legal (not a fragment). This test is passed if the FCS determines no error and the received E indicator is not set.

6. In this case, since a successful higher claim comparison has been made (figure 4.19), the receiver indicates to its transmitter to repeat the frame. It does so by setting a flag called Higher_Claim.

7. Station B then sets its T_Neg = T_Bid_Received and repeats the station A claim frame.

Station A's claim frame reaches station C which is in its T0, R0 state. Station C's state machines undergo the same sequence of operation as station B's except:

• Station C compares the T_Bid_Received (60 ms) with its own T_Req (50 ms) and finds that it has received a lower claim (two's complement numerically lower). Then it sets the L_Flag and the Lower_Claim flag indicating to the transmit state machine to strip the claim frame from station A and transmit its own claim frame.

• Station A which is continuously transmitting its claim frame receives the claim frame from station C. After the frame is received and checked for validity, station A defers (repeats C's claim frame) to the claim frame from C. It sets its T_Neg = 50 ms (T_Bid_Received), sets Higher_Claim flag, and its state machine stops transmitting its own claim frame and repeats the claim frames from C.

• At station B a similar situation occurs except that it now changes its T_Neg from 60 ms to 50 ms.

• When station C receives its own claim frame back it knows that it has won the claim process. The receive state machine sets a My_Claim flag (indicating receipt of own claim frame) for the transmitter to stop transmitting the claim frame as it has won the claim process. The transmitter then proceeds to set the T_Opr = T_Neg and resets the TRT = T_Opr. It also clears the Late_Ct. Then it issues a non-restricted token.

Note: Even though station C has won the claim process, it does not immediately begin transmitting data before issuing the token. The first token rotation is used to synchronize the various timers and flags in different stations on the network and notify all stations that the claim process has been successfully terminated.

4.10.1.2 Stage 2 – Initialization. The following sequence of figures 4.22 to 4.27 illustrate the initialization process after a station has won the bidding.

• Station A is repeating station C's claim frames when it sees the token. It knows that the claim process is complete and it initializes its timers and variables accordingly for ring operation. It sets T_Opr = T_Neg, Late_Ct = 1, and Ring_Operational = True. Late_Ct is set to one to ensure that no asynchronous data transmission occurs on the following (second) rotation. No data is transmitted on this rotation either.

• During the first rotation, all stations initialize their timers and variables to the appropriate values for ring operation.

• Meanwhile, station C is still receiving its own claim frames. It ignores the claim frames. However, if it were to receive a different claim frame after issuing the token, the claim process would be continued anew. This could occur if a new station were to join the ring just as station C won the claim process.

• Station B undergoes the same procedure as station A and repeats the token.

• When station C receives the token back, it sets its Ring_Operational flag = True and sets the Late_Ct = 1. It cannot transmit any data because the ring was not operational when the token was received. Thus, even on this second rotation, station C cannot transmit any data (synchronous or otherwise).

• It is interesting to note that other stations will be able to transmit synchronous data on the second rotation although no asynchronous transmission will be allowed because the Late_Ct = 1. At this time the TRT timer has been started in all the stations and is running at offsets of the ring latency between upstream station and downstream station.

• Station A captures the token and resets the Late_Ct = 0 and transmits the synchronous frames. The synchronous frames are typically used to deal with the allocation of bandwidth to the synchronous traffic. If no synchronous data is available for transmission, then the token is released. Asynchronous data cannot be transmitted on the first or second rotations because the Late_Ct = 1 when the token is received.

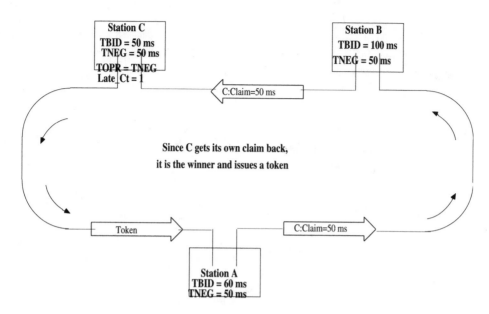

Figure 4.22   Claim process: First token rotation.

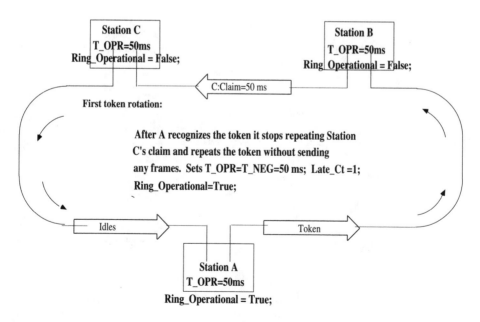

Figure 4.23   Claim process: First pass of the token.

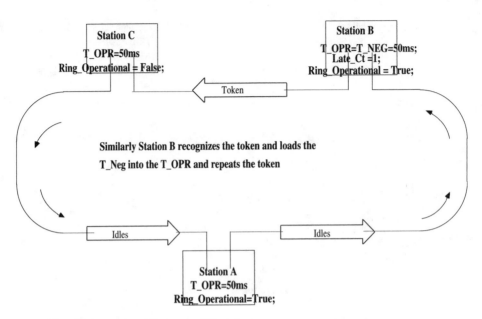

Figure 4.24 Claim process: First pass of the token.

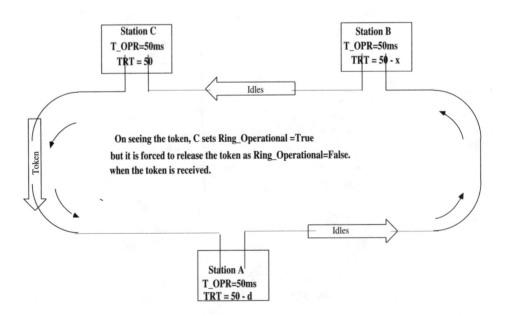

Figure 4.25 Claim process: First pass of the token.

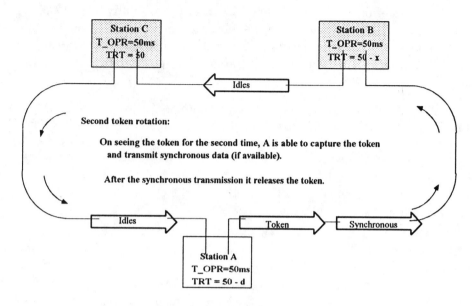

Figure 4.26   Claim process: Second pass of the token.

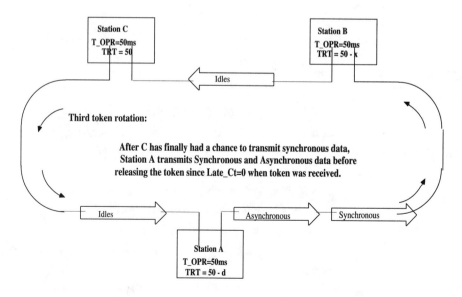

Figure 4.27   Claim process: Third token rotation.

• As can be seen from the figure 4.27, after C has finally had a chance to transmit the synchronous data (on the third rotation), station A may transmit synchronous and asynchronous data. Thereafter, all stations can transmit synchronous as well as asynchronous traffic if TRT and bandwidth allocation permit.

4.10.1.3 Claim Resolution when Multiple Stations Bid the Same TTRT. In the above examples, we saw instances of claim resolution when all stations had a different T_Bid. However, it is possible to have multiple stations bid the same T_Req. In such instances, the precedence checking followed by the MAC is:

• The bid with the fastest TTRT has precedence (i.e. numerical two's complement highest T_Bid value).

• Given equal T_Bid (e.g. 50 ms for stations B and C in figures 4.18 to 4.27), the bid with the longer address has the precedence (i.e. A 48-bit claim frame with a bid of 50 ms wins against a station issuing a 16-bit claim frame with a similar bid of 50 ms). The mechanism to distinguish long and short addresses is via the one-bit value in the frame control (FC) field. This is the L bit of the FC field. If L = 1, implies 48-bit addresses and L = 0 implies 16-bit addresses.

• Given equal T_Bid and L values (i.e. identical short or long address claim frames), the station with the highest address has precedence (i.e. the numerically highest source address value).

4.11 PERSISTENT FAULTS AND BEACON PROCESS

What happens if claim does not resolve after T_Max? Is it possible to be stuck in a fault condition? If the claim process fails to resolve, then a process known as beaconing is undertaken by the MAC. The algorithm for the beacon process is simple:

• Transmit beacons if claim fails or because of an SMT command.

• Stop transmission and go to claim if *own* beacon is received (indicated by setting of the My_Beacon flag).

Otherwise,

• Stop transmission and repeat Other_Beacon, if *another* station's beacon is received (indicated by setting of the Other_Beacon flag).

• Under all other conditions keep transmitting beacons.

The beacon process can best be explained by the figures 4.29 to 4.33. The beacon process is carried out by a specially defined MAC frame (figure 4.28). There are three conditions under which the beacon process is initiated. Different types of beacon frames are transmitted for each case.

1	1	2,6	2,6	4				2				bytes
SD	FC	DA	SA	01	02	03	04	T	E	A	C	

FC := 1L00 r010; 0xC2 (if Long address used) else 0x82 (if short
      address used); r is xmitted as 0;

DA:= Destination Adress;   DA = Null for Unsuccessful Claim;

                              DA = Multicast for Directed Beacon;

                              DA = My Long Address for Jam Beacon;

SA:= Source Address. A station must recognize a short address even
                        if it uses a long address

O1:= Octet one; 0x00 = Unsuccessful Claim;

                  0x01 = Directed Beacon;

                  0x02 = Jam Beacon;

O2 - O4:= Octet two - Octet four; 0x00; these octets may optionally be used
                                          for transmitting information;

Figure 4.28   Beacon frame content.

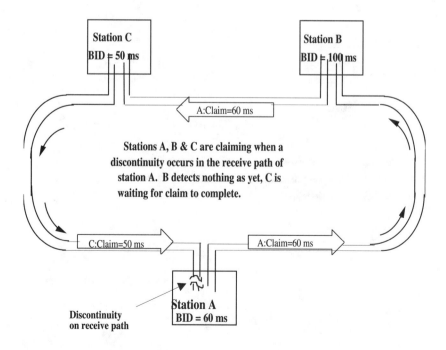

Figure 4.29   Beacon process.

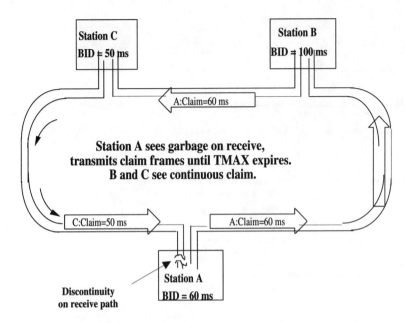

Figure 4.30   Beacon process.

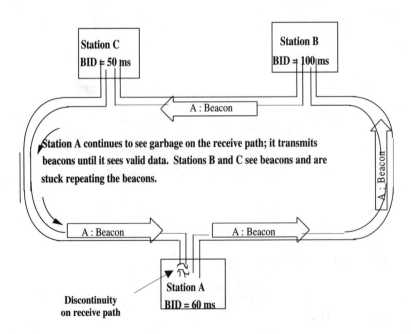

Figure 4.31   Beacon process.

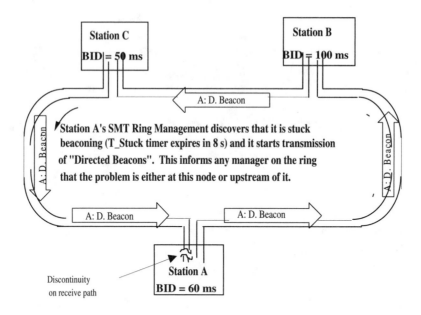

· Directed Beacon is with DA = 80:01:43:00:80:00 (MSB format)

Figure 4.32   Beacon process.

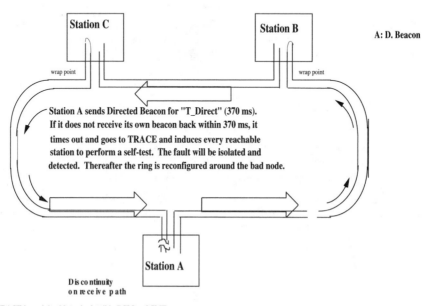

· TRACE is explained later in detail in PCM and SMT

Figure 4.33   Beacon-trace.

1. Beaconing is typically initiated when the claim process fails to complete within TMAX. This beacon is called *claim beacon*.

2. The second type of beacon is transmitted when a duplicate station is detected. In order to remove the duplicate station from the network, a process called *jamming* is initiated whereby a special beacon called *jam beacon* is transmitted. This process is explained in detail under ring management of the SMT specification.

3. If beaconing also fails to resolve, a special beacon called *directed beacon* is transmitted. This is used to convey status information to a management station on the network.

4.11.1 Duplicate Stations and Multiple Tokens

There are two possible scenarios under which more than one token can be generated:

- If there are duplicate stations on the ring *or*

- A token is created out of noise

If there are duplicate stations on the ring then it is possible that more than one station might win the claim process and issue a token. This is more likely to occur if the stations are using 16-bit addresses which are assigned by a local administrator. With 48-bit addresses, the probability of this event occurring is decreased but not eliminated. It is possible that the 48-bit address prom was programmed incorrectly or a station is using a locally assigned 48-bit address. The MAC and SMT standards have built-in feature-sets which ensure that even in the presence of a duplicate the ring becomes operational. Once the ring is operational, there are features within SMT which can detect the duplicate stations.

It is possible for some data pattern to be corrupted to a token, thus creating a duplicate token. To avoid this scenario, the symbol pattern selected for a token is the most unique pattern amongst all the five-bit symbols. The token is JK80TT. The probability of creating a duplicate token out of noise is very low (in fact less than once in a hundred years!). It would require many bits and at least two symbols to be corrupted, for a duplicate token to be generated. No other frame, besides a *null void* frame is so small. A null void frame is a frame with a start delimiter, frame control field, and ending delimiter. It has no address fields or a frame check sequence field. For a null void frame to be converted to a token, at least one bit of the FC field has to be converted from a zero to a one, and two bits of the E indicator symbol have to be converted from a zero to a one and a one to a zero respectively. The probability of this kind of noise hit is extremely low. For this and some other reasons, the null void frame which was legal in the original MAC standard has been disallowed in the MAC-2 specification.

It is also possible for a duplicate token to be generated by a faulty MAC. This situation is repeatable and easily identifiable.

4.12 BIBLIOGRAPHY

1. R. M. Grow, "A Timed-Token Protocol for Local Area Networks," Electro'82.

2. M. J. Johnson, "Proof that Timing Requirements of the FDDI Token Ring Protocol are Satisfied". *IEEE Trans. on Communications*, Vol. COM-35, no. 6, June 1987.

3. K. C. Sevcik and M. J. Johnson, "Cycle Time Properties of the FDDI Token Ring Protocol". *IEEE Trans. on Software Engineering,* Vol. SE-13, no.3, March 1987.

4. R. Jain, "Performance Analysis of FDDI Token Ring Networks: Effect of Parameters and Guidelines for Setting TTRT ". ACM SIGCOMM'90 Symposium on Communications Architectures and Protocols, Sept. 1990.

The Physical Layer

5.1 FDDI PHY Layer

In this chapter, the PHY sublayer of the physical layer is discussed in detail. The other sublayer of the physical layer is the physical media dependent (PMD) which is explained in Chapter 6. The PHY layer interacts with the MAC, PMD, and SMT layers and performs important functions such as 4B/5B encoding and decoding, elasticity buffering, smoothing, repeat filtering, clock recovery, NRZ-NRZI encoding/decoding, etc.

5.2 Physical Layer

The general function of the physical layer is to convert the data provided by the data link layer into the appropriate signal for the underlying media and vice-versa. The physical layer can be divided into two sublayers (figure 5.1):

- Physical Media Independent (PHY) and
- Physical Media Dependent (PMD)

The PHY performs functions such as data encoding, which are not related to any specific physical medium. This sublayer deals with electrical signals.

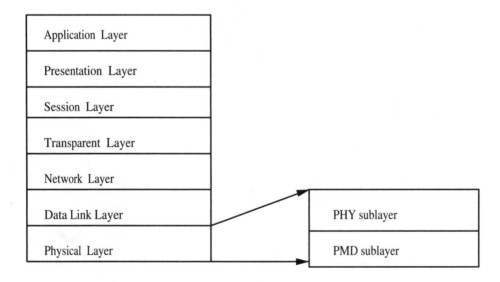

Figure 5.1 The OSI seven layer model.

The PMD layer performs media specific functions such as converting the signals provided by the PHY into signals for the specific media (e.g. converting electrical signal to optical signal) and vice-versa.

5.3   ENCODING/DECODING

One of the main functions of the physical layer is encoding and decoding of the data stream. Encoding is generally employed to:
* Provide efficient use of bandwidth
* Provide zero or minimum dc balance
* Provide maximum transitions (clock information) to synchronize receiver circuitry
* Detect errors introduced by the transmission media

There are two basic types of encoding schemes:
* Bit encoding and
* Block or group encoding

5.3.1   Bit Encoding

Bit encoding is applied to a bit stream and each bit is individually encoded. Examples of bit encoding are NRZ-L, NRZI, Manchester and Differential Manchester. Figure 5.2 shows the encoded output patterns using these methods of bit encoding for a specific input pattern.

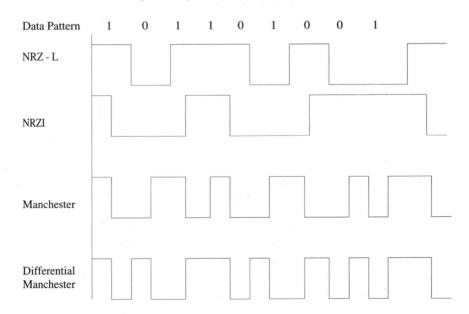

Figure 5.2   Bit encoding.

5.3.1.1 Non Return to Zero-Level (NRZ - L) Encoding. In NRZL encoding, a logical *1* is represented by a high level and a logical *0* is represented by a low level. NRZ-L is generally used to represent data in data-processing devices. The advantage of NRZ-L is that the encoding is very simple. The disadvantage is that the encoded signal may not be dc balanced which causes:

  • A need for a more complicated receiver design.
  • The absence of transitions for certain data patterns (e.g., a stream of ones or zeroes) leads to base-line wander wherein the potential difference between nodes can drift.

5.3.1.2 Non Return to Zero Invert on One's (NRZI) Encoding. In NRZI encoding, a logical *1* is represented by a transition (from its current state) during the middle of a bit interval and a logical *0* is represented by the absence of a transition. In this scheme, there can be a maximum one transition per bit interval and so a higher media bandwidth is not required. The disadvantage (as in NRZ) is that if the data is a stream of zeroes, there will be no transitions in the signal and so synchronizing the receiver circuit is more difficult. This scheme is used in FDDI and the disadvantage of the scheme is overcome by using 4B/5B encoding (discussed in section 5.3.2.1).

5.3.1.3 Manchester Encoding. The encoding rule in the Manchester scheme is as follows: A logical *1* is represented by a high to low transition in the middle of a bit interval and a logical *0* is represented by a low to high transition in the middle of a bit interval. It is evident from figure 5.2 that there is at least one transition during each bit interval. Also, there can be a maximum of two transitions per bit interval. The advantages of this scheme are:

  • Zero dc component
  • The receiver circuit can be easily synchronized to the incoming signal since there is always a transition in the middle of a bit interval.
  • The absence of a transition during a bit interval can be used to detect error conditions.

The disadvantage is that the media transmission bandwidth is doubled, since there can be two transitions per bit interval. The Manchester encoding scheme is used in the IEEE 802.3 protocol.

5.3.1.4 Differential Manchester Encoding. The encoding rule in differential Manchester scheme is as follows: A logical *1* is represented by the absence of a transition at the beginning of a bit interval and a logical *0* is represented by a transition at the beginning of a bit interval. There is always a transition in the middle of a bit interval. As in Manchester encoding, there is at least one transition per bit interval and there can be a maximum of two transitions per bit interval. This encoding scheme is used in IEEE 802.5.

5.3.2 Block Encoding

In block encoding, n bits are encoded into m bits, where m $\geqslant$ n. The fact that m is greater than or equal to n implies that the user gets to choose pattern to represent a data. Examples of block encoding are 4B/5B, 8B/10B, HDB3, 4B3T, and so forth.

5.3.2.1 4B/5B Encoding.

In 4B/5B encoding, a nibble (four bits) of data is encoded into five bits and hence it is only 80% efficient. The mapping of four bits to five bits is chosen such that:

- The number of consecutive zeros is limited to three
- Base line wander is restricted to $\pm$ 10%
- Error detection is enhanced through code violations

Implementing 4B/5B encoding is simple and straightforward (it can be implemented as a look-up table). 4B/5B may not produce a truly dc balanced output.

5.3.2.2 8B/10B.

This encoding scheme was invented and patented by IBM in 1984[1]. 8B/10B encoding provides dc component free output. 8B/10B encoding is used in ESCON™ and Fiber Channel (physical layer) protocol. 8B/10B implementation is comparatively more complex than 4B/5B.

Other block encoding schemes such as HDB3 and 4B3T are used in trunk telephone and in ISDN telephone circuits.

5.3.3 Encoding in FDDI

FDDI uses a combination of 4B/5B and NRZI encoding. NRZI was chosen to avoid a very high transmission rate. Using Manchester or differential Manchester scheme in FDDI would have led to a signaling rate of 200 Mbaud and this would have resulted in extremely high priced optical components and a very sophisticated phase locked loop (PLL) design. 4B/5B was chosen to provide minimum base line wander, high error detection and also limit the number of consecutive zeros (this overcomes the disadvantage of using only NRZI). Since 4B/5B is 80% efficient (versus 50% for Manchester), the transmission rate in FDDI is 125 Mbaud.

5.4 CODE BITS AND SYMBOLS

A code bit is the smallest signaling element used by the PHY for transmission and reception at the PHY-PMD interface. A symbol is the smallest signaling element used by the MAC layer at the MAC-PHY interface (Figure 5.3).

---

1. For technical details of 8B/10B encoding, refer to IBM Journal of Research and Development[1].

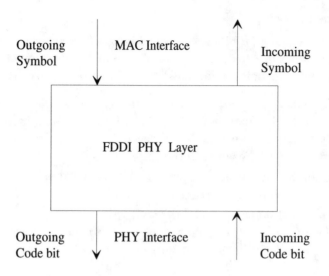

Figure 5.3 Symbol and code bit interface of PHY.

5.4.1 Code Bit

A code bit in FDDI is represented by a transition (or the absence of a transition) on the medium. The duration of each code bit is 8 ns. This is because the transmission rate is 125 Mbaud (i.e. 125 mbits/sec).

Table 5.1   Data Symbols

Symbol	Code_Group	Decimal_Value
0	11110	30
1	01001	09
2	10100	20
3	10101	21
4	01010	10
5	01011	11
6	01110	14
7	01111	15
8	10010	18
9	10011	19
A	10110	22
B	10111	23
C	11010	26
D	11011	27
E	11100	28
F	11101	29

5.4.2 Symbol

A *symbol* in FDDI is a sequence of five consecutive code bits and is also known as *code group*. A symbol is used by the MAC to represent a nibble of data. The duration of each symbol is 40 ns (five bits × 8 ns). Since each symbol consists of five bits, there are 32 possible symbols. Of these 32, 16 are used by the MAC (to represent data in nibble format). These 16 symbols are known as data symbols (see table 5.1). There are eight control symbols which are used to represent control information such as starting delimiter (SD), ending delimiter (ED), frame status (FS), and so on. (see table 5.2). Line state symbols (see section 5.5.5) are part of control symbols. The remaining eight symbols are known as violation symbols (see table 5.3). Since these symbols contain code bit patterns that violate run length (i.e., more than two consecutive zeros) or duty cycle requirement, they are called violation symbols. These symbols should not be transmitted by a node. However, a few of these symbols can be treated as halt symbols on reception as indicated in table 5.3

At the PHY-PMD interface each symbol is presented as a sequence of five bits one after the other. However, at the MAC-PHY interface each symbol is presented on a five-bit wide bus.

Table 5.2   **Control Symbols**

Symbol	Code_Group	Decimal_Group
Line State		
Q	00000	0
H	00100	04
I	11111	31
Starting Delimiter		
J	11000	24
K	10001	17
Ending Delimiter		
T	01101	13
Control Indicators		
R	00111	07
S	11001	25

Four bits represent the value of the symbol and the fifth bit (known as the control bit) indicates if it is a data or control symbol. If this bit is a *1* then the corresponding four bits represent a control symbol. If this bit is a *0*, then the corresponding four bits represent a data symbol[2].

---

2. The representation at the MAC-PHY interface is not standardized and so a PHY chip from one manufacturer and a MAC chip from another manufacturer may not be compatible.

Table 5.3   **Data Symbols**

Symbol	Code_Group	Decimal_Value
V or H	00001	1
V or H	00010	2
V	00011	3
V	00101	5
V	00110	6
V or H	01000	8
V	01100	12
V or H	10000	16

5.5 FUNCTIONAL DESCRIPTION OF PHY

The FDDI-PHY layer performs many more functions (compared to physical layers of other local area network protocols such as Ethernet and Token-Ring) such as elasticity buffering, smoothing, repeat filtering, and 4B/5B encoding/decoding. Figure 5.4 shows the functional block diagram of the PHY layer.

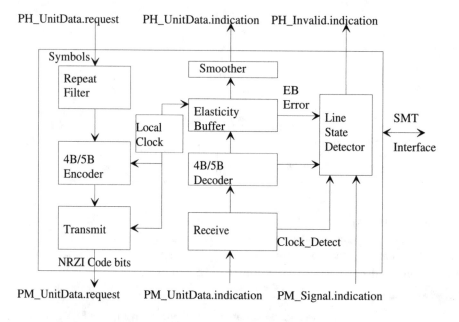

Figure 5.4   Functional block diagram of PHYsical layer.

The PHY layer has three interfaces:

• MAC interface: This is a parallel (symbol or symbol-pair wide) interface for transferring symbols between MAC and PHY entities.

• PMD interface: This is a serial (bit wide) interface for transferring NRZI code bits between PHY and PMD entities.

• SMT interface: This is a control/status interface is used by (a) SMT to control the PHY entity and (b) PHY to indicate its current status to SMT.

The individual blocks in figure 5.4 are described in detail in the following sections:

5.5.1 Receive Block

The following are the main functions performed by the receive block:

• Clock and data recovery

• NRZI to NRZ conversion

• Serial to parallel conversion and

• Framing logic (symbol alignment)

Figure 5.5 is a simple block diagram for the receive block. The bit stream in the media consists of both clock and data information and so it is necessary to separate the data from the clock before processing. The *reference clock* (figure 5.5) is generally slower than 125 MHz (ranging from 12.5 to 25 MHz) and the PLL multiplies this to a 125 MHz clock. If a PHY implementation provides the optional clock_detect signal, then the receive block will provides that signal. Also, the data from the media is in NRZI format which is converted to NRZ format before further processing. After this conversion, the serial input is

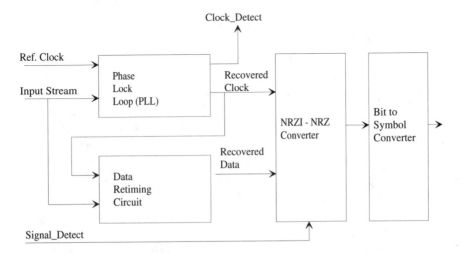

Figure 5.5   Block diagram of receive function.

converted into five-bit symbols (parallel form) before being passed on to the framing logic. The framing logic detects the presence of the starting delimiter symbols (*JK*), irrespective of the bit boundary and uses this information to align the data to the correct symbol boundary. Note that the *JK* symbol pair is unique in that it may be recognized independently of previously established symbol boundaries.

5.5.2 Elasticity Buffer

In an FDDI ring, as a bit stream enters a node, first the embedded clock information is recovered and then used to latch the incoming data. After appropriate processing, the data is latched out using a stable internal frequency source. This regeneration of data using a local clock is necessary to prevent frequency jitter from accumulating around the ring. The FDDI PHY standard requires that this local frequency be within $\pm\ 0.005\%$ of the nominal frequency which is 125 MHz. The clock regeneration process at each node prevents frequency jitter accumulation, but introduces a different problem. In each node, the recovered clock is used to latch in the data into the node and the node's local clock is used to latch the data out of the node. Since the FDDI PHY standard allows these two frequencies to differ (within a specified limit), data overflow/underflow conditions will occur unless some kind of buffering mechanism is implemented. Elasticity buffer (EB) provides this buffering mechanism to prevent loss of data.

EB is somewhat similar to a FIFO but the data is removed from the EB only after half the buffer is filled. An immediate question that comes to mind is *"What is the minimum size of this buffer and how it is derived?"*

The maximum size of a frame in FDDI ring is 4,500 bytes (which includes eight bytes of preamble). Since the main purpose is to prevent the loss of data, EB overflow and underflow should not occur when a node is receiving a maximum size frame. The EB reinitializes (recenter) to get ready to receive the next frame at the end of the preamble period. The frequency tolerance in each node is $\pm\ 0.005\%$ of the nominal frequency and therefore any two nodes can have a maximum of 0.01% difference between their local clocks.

Therefore, the minimum size of EB
= 0.01% of 4,500 bytes
= 0.01% of 45,000 bits
= 4.5 bits

Generally the EB is implemented after the serial to parallel converter and so its width is either five or ten bits (i.e. a symbol or a symbol-pair). In most VLSI implementations of the PHY, the width is ten bits and the depth ranges from three to eight. Since the buffer is reinitialized at the end of the preamble period, it is possible for the EB to insert or delete idle symbols (i.e.

the preamble symbols) depending upon the skew between the received and local clocks.

5.5.3 Smoother

Generally when a node transmits frames it is required to insert 16 idle symbols between any two frames. This is called a preamble and it is required between two frames for the following reasons:

• The MAC standard allows a MAC entity not to copy a frame with preamble shorter than 12 symbols and not repeat a frame with preamble shorter than two symbols. Also, an EB is not required to reinitialize if the preamble contains less than four symbols.

• Also, as discussed in section 5.5.2, to avoid losing data it is required to reinitialize the EB between frames.

Assuming that each node in a ring inserts enough preambles between frames, it is still possible that when frames travel from a source to a destination, the destination node may receive fewer or no preamble symbols between two frames. As discussed in section 5.5.2, this is due to the possibility of the EB in each intervening node deleting preamble symbols.

The smoother function is provided to reduce this variance in the number of preamble symbols during the frame bursts. The smoother achieves this goal by absorbing excess symbols from larger preambles and distributing them to shorter preambles. The smoother is not a symbol generator and so it needs to claim (i.e. absorb) idle symbols before it can insert (i.e. distribute) them. The smoother claims idle symbols by

• absorbing excess symbols from preambles longer than 14 symbols

• deleting symbols at the end of stripped partial frames

Examples of the smoother action are:

Input: 8 (II) Frame 6 (II) Frame IIIIII.......
Output: 8 (II) Frame 7 (II) Frame IIIIII.......

Input: 7 (II) JK FC DA SA nn 7 (II) Frame 6 (II) Frame IIIII.....
Output: 7 (II) JK FC DA SA 7 (II) Frame 7 (II) Frame IIIII......
          where *n* represents data symbols.

Note: Typically, the smoother is implemented following the block that provides the input to the line state detection logic. However, if it is placed prior to the block that provides the input to the line state detection logic, then care should be taken to prevent improper detection of:

• ILS as a result of inserting Idle symbols when not in ILS or ALS

• QLS, HLS, MLS, NLS or LSU as a result of deleting symbols that are potential noise events (see section 5.5.5.6 for definition of noise events)

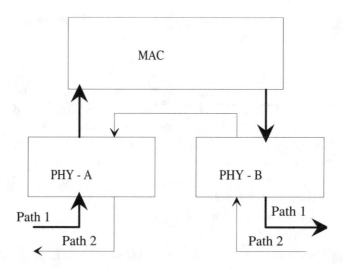

Figure 5.6 Paths with and without MAC entity.

5.5.4   Repeat Filter

The basic function of the repeat filter (RF) is to prevent code violations and invalid line states from propagating. Repeat filter is typically placed in the transmit path of PHY. The need for this function in a PHY entity can be explained by the following:

The data from the media that enters a port of a node may or may not pass through a MAC entity as in figure 5.6. In *Path 1*, there is a MAC entity and in *Path 2*, there is no MAC entity. If the symbol pattern

$$\text{II II II IV II II II......... II II II}$$

enters this node through *Path 1*, then the PHY entity will detect a Link Error event and the MAC will output the symbol pattern

$$\text{II II II II II II II........... II II II}$$

that is, the MAC entity removes the violation symbol. If the same symbol pattern entered this node through *Path 2*, then the PHY entity will detect a link error event. If the RF function is not present in the PHY entity, the same symbol pattern will leave the node. This can have a dangerous effect on the network. For example, consider a dual ring with nodes that have a MAC entity only in the primary ring as shown in figure 5.7. If the media (on the secondary path) between node 6 and node 1 corrupts the data by converting a preamble pattern II II... II into II IV.... II, then the PHY entities in the secondary path of all the nodes will increment the link error count continuously causing the link error count to exceed a pre-defined threshold which in turn will disconnect the

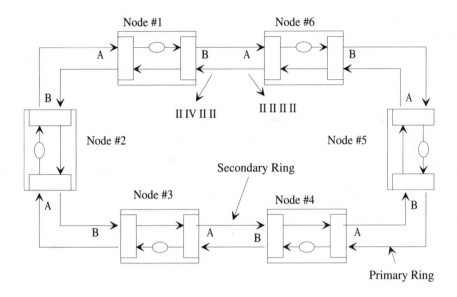

Figure 5.7   Repeat filtering.

nodes from the ring. It is very difficult for a network manager to identify the location of the origin of the problem. But if the RF function is present in a PHY then PHY-A in node 1 will detect a link rrror event and PHY-A will remove the violation symbol. Now the link error count will only be incremented in one node (i.e. in node 1) and so it is easy for a network manager to locate the origin of the problem.

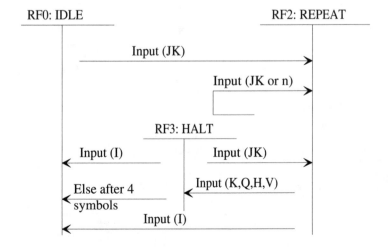

Figure 5.8   Byte-wide repeat filter state machine.

5.5.4.1 Repeat Filter State Machine. A sample implementation of Repeat Filter (symbol-pair wide) is shown in figure 5.8 in a state machine form. On reset, repeat filter (RF) will be in *idle* state (i.e. *RF0*). In this state, RF will source idle symbols until a starting delimiter (i.e. *JK*) is detected. Thus, if any idle symbol gets corrupted (as we discussed in section 5.5.3), the corrupted symbol is removed and replaced with another idle symbol. The only time the corruption is not detected is when two consecutive idle symbols get corrupted to a *JK* symbol pair. But the probability of such an occurrence is almost negligible. Once a starting delimiter is detected, the RF transitions to *repeat* state (i.e. *RF1*). In this state, the input symbol stream is repeated until a *K* (not part of a *JK*), *H, Q, V* or *I* symbol is received. If an *I* symbol is received (which, under normal conditions, indicates the end of a fragment or inter-frame gap) then the RF transitions to *idle* state. If a *K, H, Q* or *V* is received (which indicates an error condition), then the symbol is not repeated and the RF transitions to *halt* state (i.e. *RF3*). In this state, the symbol that caused this transition is changed to an *H* symbol. The next three symbols (unless *I* or *JK* symbols are encountered) are also changed to *H* symbols. The RF transitions to *idle* state if either an *I* symbol is received or after transmitting four *H* symbols from the *halt* state. If a *JK* symbol pair is received while in *halt* state, then the RF transitions to *repeat* state.

The RF state machine for a symbol wide implementation is shown in figure 5.9. While the output from a symbol wide and symbol-pair wide implementations of repeat filter may not be the same for a given input pattern, the interoperability between nodes with symbol wide and symbol-pair wide implementations is not affected.

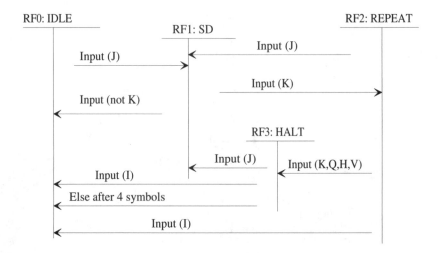

Figure 5.9 Symbol-wide repeat filter state maching.

Sample cases where the output pattern is *same* in the symbol and symbol-pair wide implementation:

Input: II II II II II IQ II II
Output: II II II II II II II II

Input: II II JK nn nn Vn nn nn nn...
          (*n* represents a data symbol)
Output: II II JK nn nn HH HH II II...

Sample cases where the output pattern is *different* in the symbol and symbol-pair wide implementation:

Input: II II II JV nn nn nn.....
Output: II II II JI II II II..... (symbol wide)
          II II II II II II II...... (symbol-pair wide)

Note: The error conditions detected by repeat filter do not include the situation in which a data symbol gets corrupted to another data symbol. Such errors will be detected by MAC and indicated as error frame (i.e. MAC increments Error_Count).

5.5.5 Line State Detection

A line state represents a longer term state of a physical link than that represented by a symbol or symbol-pair. Table 5.2 defines a set of three symbols (Q, H and I) as line state symbols. A line state is said to be entered when a certain number of these symbols are received in succession. For instance, if 16 or 17 consecutive H symbols are received, then the PHY is said to be in halt line state (HLS). When a line state is detected, the PHY entity reports to the SMT entity the occurrence of such an event. The connection management (CMT) part of the SMT uses this information while bringing up the physical connection between two adjacent nodes. The following sections explain each line state in detail.

5.5.5.1 Quiet Line State (QLS).

A PHY is said to be in quiet line state when it receives 16 or 17 consecutive quiet symbols. Also, a PHY enters QLS when the PMD indicates that the signal level on the media is below a certain threshold (i.e. in the absence of Signal_Detect). A PHY exits QLS when it receives a symbol other than a quiet symbol with Signal_Detect active. When CMT desires to transmit QLS, a PHY should send a continuous stream of Q symbols. Reception of QLS indicates one of the following:

• Absence of a physical connection
• Break in the physical connection
• A desire by the node at the other end of the connection to bring down the physical connection (permanently or temporarily).

QLS is entered in the first two scenarios due to the absence of Signal_Detect. In the third scenario, QLS is entered either due to the absence of Signal_Detect or reception of quiet symbols.

5.5.5.2 Halt Line State (HLS). A PHY enters halt line state when it receives 16 or 17 consecutive halt symbols with Signal_Detect on. HLS is exited upon reception a symbol other than H or loss of Signal_Detect. When CMT desires to transmit HLS, a PHY should send a continuous stream of H symbols. HLS is used by the physical connection management (PCM), a component of CMT, to indicate a value of 1 during the pseudo code signaling process.

5.5.5.3 Master Line State (MLS). MLS is entered upon receipt of eight or nine consecutive HQ or QH symbol pairs with Signal_Detect on. MLS is exited upon:

- receipt of a symbol pair other than HQ or QH, or
- loss of Signal_Detect

When CMT desires to transmit MLS, a PHY should send a continuous stream of HQ or QH symbol pairs. MLS is used by PCM to indicate a value of 0 to a bit during pseudo code signaling process. MLS is also used by a node to indicate a fault condition (known as *trace*).

5.5.5.4 Idle Line State (ILS). A PHY enters ILS if it receives four or five consecutive idle symbols with Signal_Detect on. ILS is exited upon:

- receipt of a symbol other than I, or
- loss of Signal_Detect

When CMT desires to transmit ILS, a PHY should send a continuous stream of idle symbols. ILS is used by PCM as delimiter between bits of information during pseudo code signaling process. ILS is also used as the preamble between MAC frames during normal ring operation. During CMT operation, ILS is detected when 16 or 17 idle symbols (this is also known as super idle line state) and during normal ring operations ILS is detected after four or five idle symbols.

5.5.5.5 Active Line State (ALS). A PHY enters ALS on reception of a JK symbol pair with Signal_Detect on. ALS is exited upon:

- loss of Signal_Detect,
- receipt of a symbol other than I, R, S, T or a data symbol, or
- satisfying the criteria for entering ILS (four idles)

If a JK symbol pair is received while in ALS, then a PHY remains in ALS. Entry into ALS indicates the start of reception of a frame or fragment.

Note: CMT cannot force a PHY into transmitting ALS. During normal ring operations, PHY will indicate only ALS and ILS to CMT.

5.5.5.6   Noise Line State (NLS). A PHY enters NLS if it detects 16 or 17 consecutive noise events without satisfying the criteria for entering any other line state. If the criteria for entering NLS and another line state are met at the same time, then a PHY enters the other line state and not NLS (i.e. other line states take precedence over NLS). NLS is exited upon meeting the criteria for entering another line state.

What is a noise event?

If a PHY decodes a Q, H, J, K or V symbol (or a symbol pair with at least one Q, H, J, K or V symbol) with Signal_Detect on, then a noise event has occurred. There are other conditions (such as EB overflow/underflow) that may also be treated as noise events. The count of noise events should be reset to zero whenever the criteria for entering another line state is met.

5.5.5.7   Unknown Line State (ULS). A PHY is said to be in a unknown line state from the time it meets the criteria to exit a line state to the time it meets the criteria to enter another line state. For instance, if the following pattern is received by a PHY, then it will be in ULS between the reception of the first H symbol and the reception of the fifteenth H symbol (and also between the first and the third I symbol). Note that on receiving the sixteenth H symbol, the criteria for entering HLS and NLS are met simultaneously. The PHY will report only HLS to the SMT.

II II II II II HH HH HH HH HH HH HH HH HH II II II II
   | ILS | ULS | HLS |ILS

5.6   PHY LAYER SERVICES

As mentioned in Section 5.5, the PHY entity interacts with MAC, PMD and SMT entities and provides a set of services for each of these interfaces as follows:

- PHY-MAC services
- PHY-PMD services
- PHY-SMT services

5.6.1   PHY-MAC Services

The PHY-MAC services consist of the following:

- Transmit_Symbol_Request
- Received_Symbol_Indication
- Invalid_Symbol_Indication

5.6.1.1   Transmit_Symbol_Request. This primitive is used by a MAC (or another PHY) entity to transfer a symbol (or a symbol pair) to a PHY entity. When a MAC uses this primitive, the symbol that is transferred is generally one of the following:

J, K, I, n, R, S and T.

If another PHY entity uses this primitive, then the symbol that is transferred is one of the following: J, K, I, n, R, S, T and H. A PHY entity can also, optionally make a request to transfer either a Q or V symbol. The ISO standard document identifies this primitive as *PH_UNITDATA.request*.

5.6.1.2 Received_Symbol_Indication. This primitive is used by a PHY entity to transfer a symbol to a MAC (or another PHY) entity. The symbol that is transferred is one of the following: J, K, I, n, R, S, T and H. A PHY entity can optionally also transfer Q or V symbols. The ISO PHY standard document identifies this primitive as *PH_UNITDATA.indication*.

5.6.1.3 Invalid_Symbol_Indication. A PHY entity uses this primitive to indicate:

- QLS, HLS, MLS or NLS detection
- Elasticity buffer overflow/underflow condition
- Deassertion of Signal_Detect signal

The ISO PHY standard document identifies this primitive as *PH_INVALID.indication*.

5.6.2 PHY-PMD Services

The PHY-PMD services consists of the following primitives:

- Transmit_Bit_Request
- Received_Bit_Indication
- Received_Signal_Indication

5.6.2.1 Transmit_Bit_Request. This primitive is used by a PHY entity to transfer a NRZI code bit to PMD entity. The ISO PHY standard document defines this primitive as *PM_UNITDATA.request*.

5.6.2.2 Received_Bit_Indication. This primitive is used by a PMD entity to transfer a NRZI code bit to PHY entity. The ISO PHY standard document defines this primitive as *PM_UNITDATA.indication*.

5.6.2.3 Received_Signal_Indication. This primitive is used by a PMD entity to indicate to a PHY entity the level of the optical signal received by the PMD entity. The status is true if the signal level is above a minimum threshold value and false if less than the threshold value. The ISO PHYstandard document defines this primitive as *PM_SIGNAL.indication*.

5.6.3 PHY-SMT Services

The PHY-SMT services consist of the following primitives:

- Transmit_Linestate_Request
- Receive_Status_Indication
- Control_Status_Request

5.6.3.1 Transmit_Linestate_Request. This primitive is used by SMT entity to request a PHY entity to transmit a given line state. A PHY entity should have the capability of transmitting quiet, halt, master and idle line states. When SMT requests a PHY entity to transmit one of the above line states, PDUs from a MAC or symbols from another PHY entity are ignored. This primitive is used by CMT portion of SMT and takes precedence over Transmit_Symbol_Request (defined in section 5.6.1.1). The ISO PHY standard document defines this primitive as *SM_PH_LINE-STATE.request.*

5.6.3.2 Receive_Status_Indication. This primitive is used by a PHY entity to indicate to SMT entity of the following:

- Type of line state received
- Detection of EB overflow/underflow

The ISO PHY standard document defines this primitive as *SM_PH_STA-TUS.indication.*

5.6.3.3 Control_Status_Request. This primitive is used by SMT entity to make request to a PHY entity to perform certain operations. Few of the operations are as follows:

- Reset a PHY entity (such as initialize various state machines in a PHY entity)
- Place (and remove) a PHY entity in a loop-back configuration
- Obtain certain status information

The ISO PHY standard document defines this primitive as *SM_PH_CONTROL.request.*

5.7 BIBILIOGRAPHY

1. *IBM Journal of Research and Development*, Volume 27, No. 5, September 1983 issue.
2. K. Annamalai, "FDDI Physical Layer Implementation Considerations," *SPIE*, Vol. 991, Fiber Optic Datacom and Computer Networks, 1988.
3. Raj Jain, "Error Characteristics of FDDI," *IEEE Transactions on Communications*, Vol. 38, No. 8, August 1990.

Physical Media Dependent (PMD) Layer

6.1 INTRODUCTION

The PMD layer (lower sublayer of the physical layer) provides the media interface specification for the transceivers, optional optical bypass switch and the media interface connector (MIC). The PMD layer interacts with PHY and SMT layer (figure 6.1). This chapter describes the PMD components and the different media choices.

6.2 PMD COMPONENTS

The basic function of the PMD is to convert the signal provided by the PHY layer to a form that is suitable for the underlying media. For example, when fiber is used as the medium of transmission, the PMD converts the electrical signal provided by the PHY layer into optical signal for transmission and the optical signal from the media into electrical signal for the PHY layer.

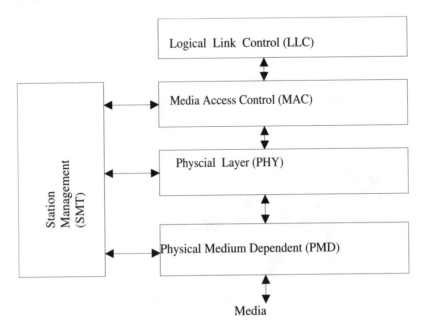

Figure 6.1   Physical media dependent (PMD) layer.

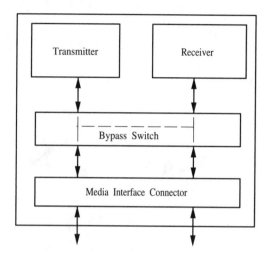

Figure 6.2 Components of PMD.

The different components of the PMD layer are shown in figure 6.2. Initially, we will assume that fiber is the media for transmission. In the latter part of this chapter, we will discuss the issues related to alternative media such as low cost fiber (LCF), shielded twisted pair (STP) and unshielded twisted pair (UTP).

6.2.1 Optical Transmitter

An optical transmitter converts electrical signal to optical signal. Two common types of optical transmitters are:
- Light emitting diode (LED)
- Laser diode

These are commonly used because they are made of semi-conducting materials which require lower power than gas lasers such as helium-neon lasers and the smaller size makes interfacing with smaller diameter fiber easier. An optical transmitter is usually specified by the following parameters:
- center wavelength
- spectral width
- average power

Center Wavelength is the *desired* wavelength of the optical signal produced by a transmitter. Theoretically, a fiber optic transmitter's attenuation and dispersion characteristics are best at 1,550 nm, 1,300 nm or 850 nm (figures 6.3a and b). Optimum performance would be obtained if a transmitter emitted light at such a single wavelength. However, the transmitters available today are not single wavelength sources but transmit across a wave-

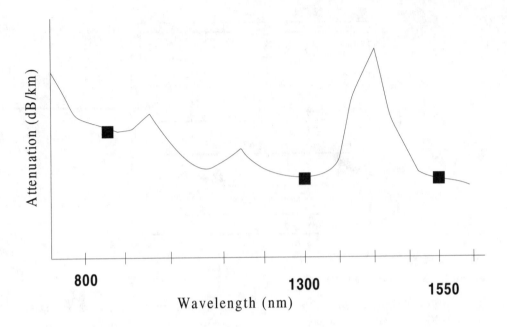

Figure 6.3a   Spectral attenuation in glass fibers.

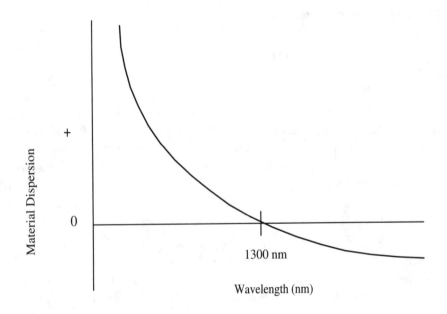

Figure 6.3b   Wavelength vs. material dispersion.

length spectrum *centered* around the *desired* wavelength. Thus in FDDI the center wavelength is in the range of 1,270 to 1,380 nm. Transmitting at 1,550 nm, 1,300 nm and 850 nm wavelengths is advantageous because for a given distance less power is required than would be at other wavelengths. Typically, LEDs are used as optical transmitters at 850 and 1,300 nm wavelengths and laser diodes are used as optical transmitters at 1,550 nm wavelength. In FDDI, the 1,270 to 1,380 nm range is used for two reasons:

 • LEDs are cheaper than laser diodes.

 • Modal dispersion is higher at 850 nm than at 1,300 nm. Thus, better bandwidth distance product is obtained.

Spectral Width is a measure of different wavelengths produced by a transmitter. Spectral width can be defined as the difference in wavelength between two points (one on each side of the center wavelength) in a plot of power versus wavelength. The signal power is maximum at the center wavelength and decreases on either side. The difference in wavelengths between the points at which the signal power is half the power of the center wavelength is called full width half maximum (FWHM). The FWHM is the spectral width of a source. Figure 6.4 shows the spectral width for a typical LED and

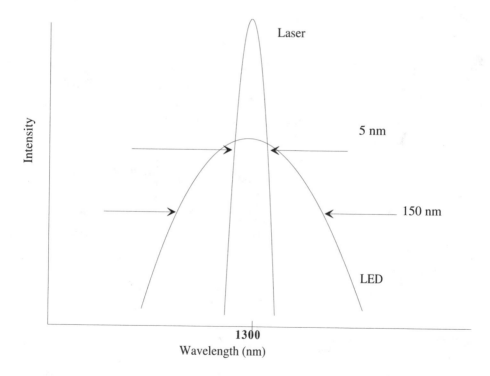

Figure 6.4   Spectral width.

laser. The smaller the spectral width, the more coherent is the source. laser diodes are more coherent than LEDs. A transmitter which produces a single wavelength optical signal is a monochromatic source of light.

Average Power is the mean of the minimum and maximum level of power output from a given transmitter and is measured in dBm. It is typically specified for a given fiber core-size and numerical aperture. Average power for LEDs ranges from -10 dBm to -30 dBm and for lasers the range is from -3 dBm to $+1.0$ dBm. Higher the average power level, higher the loss that can be tolerated due to media, connectors, splices, and so on.

6.2.1.1 Light Emitting Diode.

LED is a p-n junction semiconductor diode. The n-type semiconductor has a number of free electrons and the p-type semiconductor has a number of free holes (the absence of an electron is a hole). When a p-type and n-type semi conductor material are fused an energy barrier is produced at the boundary. Free electrons in the n-type semiconductor do not have enough energy to cross the barrier and enter the holes in the p-type semiconductor. The same is true for holes.

When a forward bias (as shown in figure 6.5) is applied to a LED, the energy barrier is reduced. As a result, electrons move into holes and optical energy is released in the process. The wavelength of the optical signal generated is governed by the following equation:

$$\lambda = 1.24 / W_g$$

where W_g is the energy gap between upper energy level (*conduction band*) and lower energy level (*covalence band*). The different semiconductor materials that are used in LED are GaAs, AlGaAs, InGaAs and InGaAsP. The mean time between failure (MTBF) for an LED is generally higher than that for a laser. There are two types of LEDs: surface emitting LED and edge emitting LED. Figure 6.6 shows a surface emitting LED and figure 6.7 shows an edge emitting LED. Table 6.1 shows the main difference between these types of LEDs.

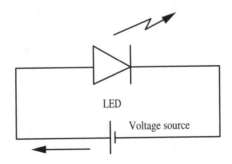

Figure 6.5 Forward bias circuit for LED.

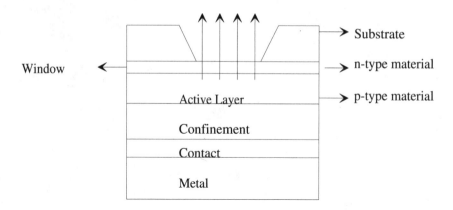

Figure 6.6   Surface emitting LED.

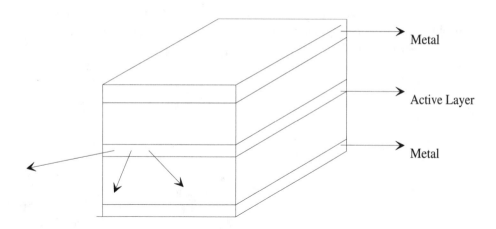

Figure 6.7   Edge emitting LED.

Table 6.1   **Surface Emitting LED vs. Edge Emitting LED**

Surface Emitting LED	Edge Emitting LED
Wider spectral width	Narrower spectral width
Smaller rise time	Larger rise time
Lower power	Higher power

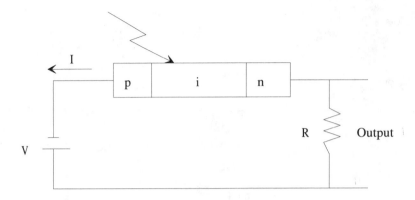

Figure 6.8 Reverse biased PIN photodiode.

6.2.1.2 Laser Diode. The principle of operation and structure of a laser diode is similar to that of an LED. However, there are several differences between LED and a laser diode.

The width of p and n type material is very small in laser diodes. For example, the width is around 20 m in LED and around 1 m in laser diodes. Laser diodes are typically edge emitting type and are much more temperature sensitive than LEDs (i.e. for a given bias current, with an increase in temperature the reduction in output power is larger in laser diodes than in LED). The spectral width of a laser diode is smaller than that of a LED (figure 6.4).

6.2.2 Optical Receiver

The two popular types of optical receivers are positive intrinsic negative (PIN) photodiode and avalanche photodiode (APD). Both are semiconductor type detectors. Of these two types, PIN photodiode is cheaper and more commonly used than APD. A receiving circuit using a PIN photodiode and an APD are shown in figures 6.8 and 6.9 respectively. Gain provided by a PIN photo-

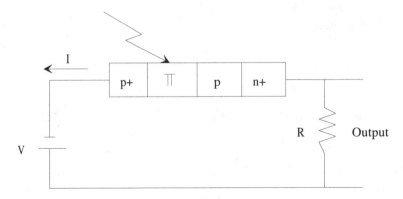

Figure 6.9 Reverse biased avalanche photodiode.

diode is smaller than that provided by an APD. When light strikes photons are absorbed in the depletion region. This leads to creation of electrons and holes. Larger electrical forces in the depletion region cause these holes and electrons to accelerate. When these fast moving charges collide with neutral atoms, they create more electrons and holes. These newly created charges, in turn, create more electrons and holes. This avalanche multiplication process results in more internal gain. Since the gain in internal (as opposed to external gain with PIN photodiode), the signal to noise ratio is higher. The disadvantage with APD is that it requires a large reverse bias voltage. Sensitivity of a PIN diode receiver is more than sufficient to meet the FDDI specifications (see table 6.4). Also, the PIN diodes exhibit acceptable signal to noise ratio at these power levels.

6.2.3   Optical Bypass Switch

In a dual ring topology, if a node is removed from the network, the adjacent nodes will wrap and the dual ring becomes a single ring (figure 6.10). If one more node (other than the wrapped ones) is removed, the single ring is segmented into two rings (figure 6.11). If the nodes that were removed were equipped with an optical bypass switch, wrapping and segmentation of the ring could have been prevented. Optical bypasses are useful in preventing

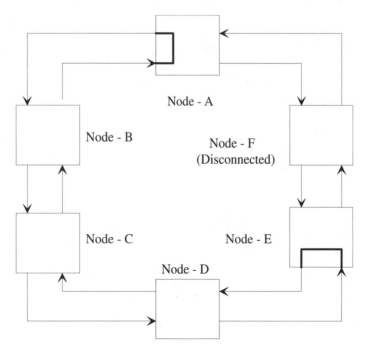

Figure 6.10   Wrapped ring (due to single disconnected node).

ring wraps due to node deinsertions. However, they cannot prevent ring wraps and segmentation due to physical cable faults or breaks. In FDDI, it is optional for a node to employ optical bypass switch. Care should be taken in selecting an optical bypass switch since, in bypassed state, optical power from transmitter is directly fed back to the receiver through the bypass switch. If the bypass switch does not attenuate the optical signal, the receiver may get saturated and in some cases, the receiver may even be damaged. There are different types of optical bypass switches such as moving fiber and moving mirror or moving prism. A moving mirror type of optical bypass switch is shown in figure 6.12a and b. Moving fiber type is more commonly used because it provides better insertion loss and is more reliable. There is no need for an optical bypass switch in a single attachment station since concentrators to which it is connected will perform the function of bypassing the node.

When the optical signal traverses through the optical bypass switch, it gets attenuated. The maximum attenuation that is allowed (in FDDI) is 2.5 dB. The typical attenuation in an optical bypass switch is approximately 1.25 dB. Also, the amount of time taken by an optical bypass switch to transition from normal state to bypassed state (and vice-versa) should not be more than 25 ms. Table 6.2 indicates the specifications for optical bypass switches.

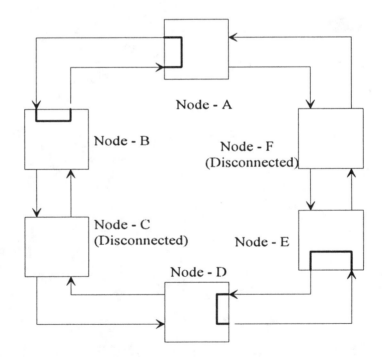

Figure 6.11   Segmented ringe (due to two disconnected nodes).

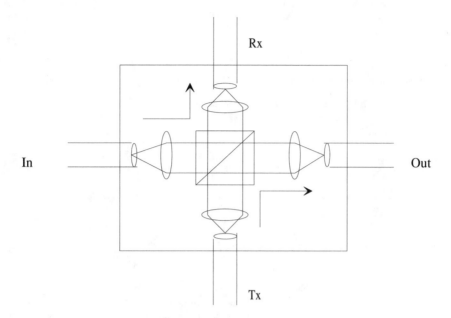

Figure 6.12a   Moving prism bypass switch–normal state.

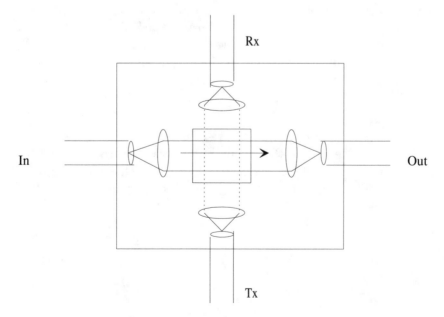

Figure 6.12b   Moving prism bypass switch–bypassed state.

Figure 6.13   ST connector.

Table 6.2   **Optical Bypass Switch Specifications**

Characteristics	Minimum Value	Maximum Value
Attenuation (dB)	0.0	2.5
Interchannel Isolation (dB)	40.0	n/a
Switching Time (ms)	n/a	25.0
Media Interruption (ms)	n/a	15.0

6.2.4   Media Interface Connector (MIC)

Media interface connector connects a node to the medium. There are three different types of connectors. They are ST connector, SC connector and MIC connector (a fixed shroud duplex connector). ST and SC connectors are shown in figures 6.13 and 6.14 respectively. Both ST and SC are non-FDDI standard connectors. However, most of the first generation FDDI products used the ST connector exclusively. The FDDI standard MIC connector plug is shown in figure 6.15. In FDDI, certain port combinations are not allowed. For example, an M port should not be connected to another M port. To prevent these kinds of connections, the connectors are *keyed*. Figures 6.16 and 6.17 show the MIC receptacle keying for different types of ports for a multimode fiber and single mode fiber systems respectively. Tables 6.3 and 6.4 list the specifications for the optical signal at the output and the input interface of a MIC receptacle for multimode fiber systems.

6.3   MEDIA

During the initial development of FDDI, the underlying assumption was that the fiber cable will be the medium of choice for transmission. However, efforts are now underway to evaluate copper as an alternate for fiber as the medium of transmission. First fiber is discussed in detail and then the other media choices are discussed.

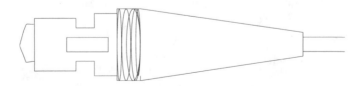

Figure 6.14   SC connector.

Table 6.3   **Specifications for the Optical Signal at the Output of MMF-MIC**

Characteristics	Minimum Value	Maximum Value
Center Wavelength (nm)	1270	1380
Average Power (dBm)	-20.0	-14.0
Rise Time (ns, 10-90%)	0.6	3.5
Fall Time (ns, 90-10%)	0.6	3.5
Duty Cycle Distortion (ns, peak-peak)	0.0	1.0
Data Dependent Jitter (ns, peak-peak)	0.0	0.6
Random Jitter (ns, peak-peak)	0.0	0.76
Extinction Ratio (%)	0.0	10.0

Table 6.4   **Specifications for the Optical Signal at the Input of MMF-MIC**

Characteristics	Minimum Value	Maximum Value
Center Wavelength (nm)	1,270	1,380
Average Power (dBm)	-31.0	-14.0
Rise Time (ns, 10-90%)	0.6	5.0
Fall Time (ns, 90-10%)	0.6	5.0
Duty Cycle Distortion (ns, peak-peak)	0.0	1.0
Data Dependent Jitter (ns, peak-peak)	0.0	1.2
Random Jitter (ns, peak-peak)	0.0	0.76

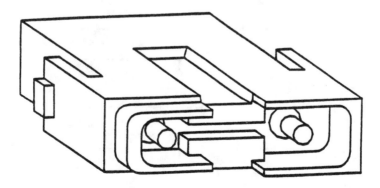

Figure 6.15   Media interface connector (MIC) plug.

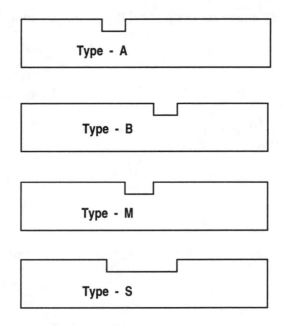

Figure 6.16 MMF–MIC receptacle keying.

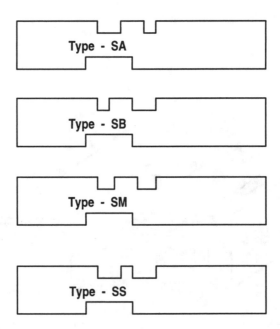

Figure 6.17 SMF–MIC receptacle keying.

6.3.1   Optical Fiber

FDDI is a high speed network and the distance between any two adjacent nodes can be up to 2 km. Fiber is initially chosen as the medium for FDDI networks because of the following characteristics:

High Capacity.   In digital fiber systems (such as FDDI), bandwidth is measured as a product of frequency and distances. This is so because the longer the distance, the greater is the distortion. Fiber optic systems have bandwidths of hundreds of MHz over distances of tens of kilometers.

Low Attenuation.   Fiber attenuates signals by very small amounts. Since the attenuation is very low, lesser number of repeaters are required when the distance between the source and destination is very large.

Electromagnetic Interference.   Optical signals are not affected by external electromagnetic fields. Also, they are immune to radio frequency interference (RFI). Since, the optical signals do not radiate any form energy other than optical, they do not affect any other equipment.

Lightweight.   Fiber cables are very light and small in size. So a large amount of fiber cables occupies less space than copper cables.

Security.   It is very difficult to tap signals from a fiber cable. Any tap results in a significant and noticeable drop in the signal level, and with the appropriate equipment, the tap can be easily located.

6.3.1.1   Fiber Structure.   Figure 6.18 shows the structure of a fiber cable. The central-most part of the cable (known as the core) is the conductive part of the cable (i.e. the light propagates through this portion). The core is made of either glass or plastic. The part that surrounds the core is known as cladding. Cladding can also be either glass or plastic. But the refractive index

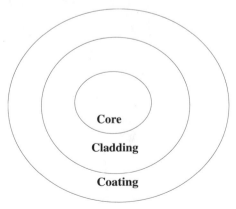

Figure 6.18   Structure of a fiber.

of the cladding is lower than that of the core so that the light waves are reflected by the cladding back into the core. Generally the difference in the value of the refractive index of the core and that of the cladding is less than 2%. A plastic sheath surrounds the cladding and it provides the necessary support for the cable. It also minimizes the bending effect. Typically, the size of a fiber is expressed as "core diameter / cladding diameter" (e.g. 62.5/125 m).

6.3.1.2   Types of Fiber.   There are two basic types of fiber:
- Multimode fiber (MMF)
- Single mode fiber (SMF)

6.3.1.2.1.   Multimode Fiber.   In multimode fiber, the core size varies from 50 to 200 micron and the cladding diameter ranges from 125 to 230 micron. In MMF the light from a source travels along different paths in the fiber. Since optical signal travels along different paths from one end to another, different signals arrive at different times resulting in the dispersion of the signal. There are two types of multimode fiber:

- Step index fiber
- Graded index fiber

Step Index Multimode Fiber.   In step index multimode fiber, the refractive index of the cladding is smaller than that of the core and the refractive index is the same for the entire part of the core (figure 6.19). In this type of fiber, since multiple propagation paths exist and each path is of different length, the signals along each of these paths take a different amount of time to propagate (figure 6.20) and hence "spreading" of signal (known as dispersion) occurs.

Graded Index Multimode Fiber.   In graded index multimode fiber, the refractive index of the core varies as shown in figure 6.21. Due to this kind of variation of refractive index of the core, more amount of light (compared to step index multimode fiber) arrives at the same time at the receiving end of

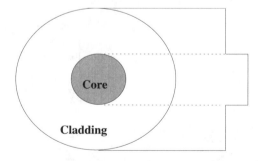

Figure 6.19   Variation of refractive index in a step-index multimode fiber.

the fiber (figure 6.22). Because of this, the dispersion is smaller compared to step index multimode fiber.

The attenuation and the dispersion characteristics of graded index MMF is intermediate between step index multimode and single mode fiber. The following are the commonly used graded index multimode fibers:

- 62.5/125 μm
- 85/125 μm
- 100/140 μm

6.3.1.2.2.   Single Mode Fiber.   In single mode fiber, the core size is very small (comparable to the wavelength of the optical signal carried) and so light passes along the axis of the cable. Because of this the attenuation of light signal is very small and the dispersion is minimal. Hence, single mode fiber can be used to carry an optical signal through larger distances. Generally, single mode fibers are the step index type. Typical size of a single mode fiber is 10/125 micron. Table 6.5 describes the differences between single and multimode fibers systems.

Table 6.5   **Single Mode System vs. Multimode Systems**

Single Mode System	Multimode System
Single mode system	Multimode system
Uses single mode fiber	Uses multimode fiber
Laser transmitter	LED transmitter
APD detector	PIN or APD detector
Higher bandwidth and longer distances (~ 100 Gb-km)	Lower bandwidth and shorter distances (~ 1 Gb-km)
More Expensive	Less Expensive

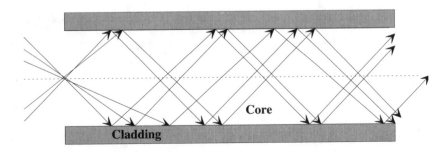

Figure 6.20   Path of optical rays in a step-index fiber.

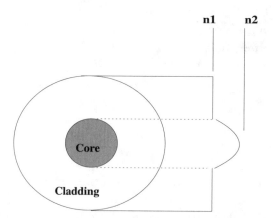

Figure 6.21 Variation of refractive index in a graded-index multimode fiber.

6.3.1.3 Fiber Care and Handling. Fiber cables should be handled as indicated below:

- Always close the ends of a fiber cable with a dust cap
- Keep the mating portion of a fiber cable clean (isopropyl alcohol can be used for cleaning purposes)
- Do not pull fiber too hard
- Do not tie fiber cable into knots
- Never look directly into an active fiber optic cable

6.3.2 Copper Media

Initially, only fiber was chosen as the media for FDDI networks for various good reasons. However, bringing FDDI to the desk-top market (with fiber as the only media) is very difficult because of the following reasons:

- Most of the existing buildings are already wired with copper cable (copper cabling accounts for 95% and fiber cabling accounts for the remaining 5%)

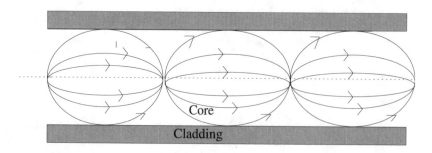

Figure 6.22 Path of optical rays in a graded-index fiber.

• Fiber optic transceivers are still expensive (though the price is gradually decreasing) and they cost as much as (or more than) the MAC and PHY chips (and in some cases the system interface) put together.

• Fiber cable is not expensive (fiber optic cable price is comparable to that of copper) but installation of fiber cable is very expensive. Copper cable installation cost is around $2.75/meter whereas fiber cable installation cost is around $7.75/meter.

While fiber is the best choice (and in most cases the only choice) for a backbone FDDI network, it is not necessary to use fiber for the tree structure (front-end desktop connectivity) of a FDDI network since the distance between a closet and a desk-top unit is generally less than 100 meters. It is possible to use different cable types (fiber or copper) in any segment of a FDDI network so long as they meet certain critical FDDI performance parameters such as bit error rate.

6.3.2.1 Issues in Using Copper Rather than Fiber.

Fiber is an excellent medium for high speed transmissions. It does not suffer from electro-magnetic emissions, it is not susceptible to noise and crosstalk, and it has very low attenuation. Copper suffers from the all these limitations. Depending on the type of copper cable the noise susceptibilities and emissions can vary significantly.

Susceptibility is the phenomenon by which external noise signals are coupled into the cable thus distorting the carrier signal. This is known as electro-mechanical/radio frequency interference (EMI/RFI) if the noise sources are external noises from electrical motors, television signals, radio frequencies, microwave and so on. Usually noise picked up from adjacent pairs of cable is referred to as crosstalk although crosstalk is EMI/RFI.

Emission is the phenomenon by which the copper cable radiates the signal rather than propagating the signal along the cable. This is what induces crosstalk. Crosstalk is a form of noise and noise can be overcome by increasing the signal strength. The problem is that the FCC has strict laws for emissions at frequencies greater than 30 MHz and increased signal strength also increases the emissions. Therefore what was not a problem with Ethernet (or a lesser one) becomes a bigger problem with FDDI.

Attenuation is the loss of signal amplitude as it travels along a cable. In copper wiring, the higher the frequency, the higher the attenuation and hence the smaller the maximum distance. When using copper cables for high-speed networking, the distances targeted are typically 100 m or less which is the distance from the desktop to the closet. Attenuation can be offset by increasing the signal strength. This would, however, lead to greater emissions and crosstalk. Hence, the use of copper cabling is not being considered for backbone networks where distances are typically in the 500 m to 2,000 m range.

There are different types of copper cables. The three major types are:

- Coaxial cable
- Shielded twisted pair (STP)
- Unshielded twisted pair (UTP)

6.3.2.2 Coaxial Cable. Coaxial cable is a two-conductor cable used to transmit both analog and digital signals. It offers substantially larger bandwidths than twisted-pair wire and can support high data rates with high immunity to electrical noise and good noise characteristics. A coaxial cable consists of a central carrier wire which is surrounded by an outer conductor. The inner conductor can be either solid or stranded, and the outer conductor can be solid or braided too. The inner and outer conductors are isolated by a solid dielectric. The outer conductor is sheathed in a plastic jacket.

Coaxial Cable Applications. Coaxial cable is used for the following applications:

- Distribution of television signals to homes (also coaxial cable is used to connect TV to the connectors on the wall)
- Local area networks (Ethernet and Token-Ring network protocols use coaxial cables).

In the LAN market, coaxial cables are no longer used in new installations and have been replaced with the twisted pair cables due to ease of use and lower cost. Coaxial cables are bulky and require a crude tap mechanism in order to access the cable from the desktop. This type of cable is not being considered for use in FDDI although it meets the FDDI bandwidth and error characteristics requirements and has the lowest emissions, noise susceptibility and attenuation of the three common types of copper cables.

6.3.2.3 Twisted Pair Cables. Twisted pair wiring has been traditionally used in the telephone industry where it has been used for well over seventy years. Twisted pair wire has two insulated wires that are twisted around each other. The twisted-pair cables come in various gauges (diameter), different twists, and different insulating materials. There are two basic types of twisted pair cables:

- Shielded twisted pair
- Unshielded twisted pair

Shielded Twisted Pair (STP). Shielded twisted pair (STP) cables consist of insulated conductors with each pair being individually shielded. The shielding is made of braided copper and/or a spirally wrapped metal foil. One-hundred-fifty-ohm impedance STP cables are largely used in Token-Ring networks and to a lesser degree in Ethernet and FDDI networks. These cables provide better noise immunity and have lower emissions than Unshielded Twisted Pair (UTP) cables. These cables are more expensive to install than UTP and constitute a very small percentage of the installed base of copper wiring. STP is used in Token-Ring environments and in noisy workshop floor type environments. Several forums have released specifications for FDDI over STP. The most prominent amongst them are *The Green Book* released by DEC, AMD, Motorola, SynOptics and Chipcom, and *SDDI* by IBM. Several vendors already supply FDDI equipment with STP interfaces compliant with one of the above two specifications.

Unshielded Twister Pair (UTP). Unshielded twisted pair (UTP) cables consist of insulated conductors formed into individually twisted pairs and enclosed by a shielding jacket. There are several categories of UTP cables which are categorized by the Electronics Industries Association/Telecommunications Industries Association (EIA/TIA) TR 41.8.1 committee. Two of the most popular types of UTP cables are:

• Data grade UTP. It is also referred to by its EIA/TIA name which is *category 5*.

• Voice grade UTP. It is also referred to by its EIA/TIA name which is *category 3*. IBM refers to it as Type 3 cable for its 4 Mbps Token-Ring. This is the more popular telephone cable.

Generally four pairs of UTP wires are bundled in the same cable. Other combinations involve eight and twenty-five pair bundles. The major difference between the category 5 (data grade) and category 3 (voice grade) type of UTP cables is in the number of twists per inch and the uniformity of the twists. Voice grade UTP cable has approximately two 5-twists per foot, and Data grade UTP cable has two 5-twists per inch. The twist cause the electromagnetic fields of the two wires which are in opposite directions (differential signals) to cancel each other so that the net cable radiation is zero. This also minimizes crosstalk coupling into other pairs of wire within the same bundle. The greater the number of twists per inch the lower the radiation and crosstalk.

UTP cables are used in Ethernet, Token-Ring and ARCnet. FDDI over UTP is being considered within the ANSI X3T9.5 committee.

Issues in Transmitting 100 Mbps over UTP Cable. There are many advantages to implementing FDDI over copper cable, not the least of which is to utilize the enormous installed base of UTP cables to the desktop. However there are many issues in high speed transmissions over UTP. Some of the issues are:

• UTP category 3 cable have not been characterized beyond 16 MHz frequency. Since FDDI data transmissions are at 100 Mbps it is difficult to predict the noise and attenuation characteristics of category 3 at 100 Mbps.

• The widely installed telephone cable is not always category 3. Often it is category 2 which is installed. Category 2 has even worse EM Interference and Radiation, crosstalk and attenuation characteristics than category 3.

• The UTP cables are generally installed in bundles of four or eight pairs. In some installations, two pairs are used for data communications and two pairs are used for telephony. The presence of additional signals (disturbers) in the same bundle causes severe crosstalk in category 3 cables at FDDI speeds. Even with 10BaseT Ethernet many vendors strongly recommend against having multiple transmissions within the same bundle.

• The Federal Communications Commission (FCC) has stringent regulations for electro-magnetic emissions beyond the 30 MHz frequency spectrum. This is especially true for what is called FCC Class B certification which is for home or commercial use. FCC Class A certification is for commercial use only and has less stringent requirements than FCC Class B.

• The category 5 cable has superior noise immunity and emissions capability. It can carry 100 MHz frequencies for up to 100 m and meet the FDDI bit error rate (BER) requirements.

• Poorly designed cross-connects (or punch-down blocks) for terminating desktop cable runs can cause significant reflections and attenuation. The losses become glaring at the higher speeds.

• Poorly terminated connectors can also cause significant losses.

Despite these technical and implementation issues the ANSI X3T9.5 committee is close to resolution on a scheme for transmitting 100 Mbps FDDI over UTP category 5 cable.

ANSI X3T9.5 (FDDI) Proposal for 100 Mbps over UTP
Category 5. The ANSI committee is developing a 100 Mbps FDDI over two pairs of category 5 cable for 100 m distances. In order to achieve FCC Class A (and preferably FCC Class B) certification, the committee has decided to use a different encoding scheme Multi-Level Transmit-3 (MLT-3) and a stream cipher scrambler.

The MLT-3 reduces the frequency spectrum (see figure 6.23) and the stream cipher spreads the energy spectrum to reduce emissions and interference issues. Although work remains to be done proprietary (albeit using proposed standards schemes), solutions have been developed and FDDI adapters with UTP category 5 PMD are available in the market. Figures 6.24a and 6.24b show a SAS implementation using a parallel and a serial scrambler along with MLT-3 encoding/decoding respectively.

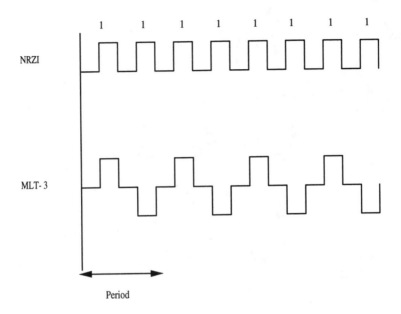

Figure 6.23   MLT-3 encoding.

6.3.2.4   Low Cost Fiber (LCF).   For those who need the performance of fiber to the desktop at a price comparable to copper cabling, the ANSI X3T9.5 committee has developed a low cost fiber PMD. The new low-cost PMD targets reduced maximum distances (500 m versus the original 2 km). This reduces the cost of the transceivers because the transmitter can be less powerful and the receiver sensitivity can be relaxed while maintaining the same cable plant.

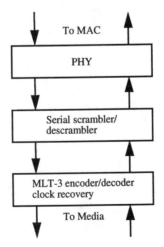

Figure 6.24a   Copper PMD using serial scrambler and MLT-3.

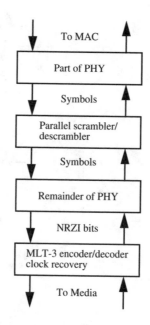

Figure 6.24b   Copper PMD using parallel scrambler and MLT-3.

CHAPTER 7

Station Management –

Link and Node Management

7.1 INTRODUCTION

Station management (SMT) is the fourth specification of the four-part FDDI set of standards. FDDI is the only LAN to have management protocols specifically designed for the FDDI subnetwork characteristics. Unlike other LANs where management capabilities were added after the standard was developed, FDDI has a sophisticated, built-in network monitoring and management capability.

SMT is unique in more than one way. It is the collective work of a large number of companies (and organizations), which may be considered by some to be a dubious distinction because of the number of years it took to develop. It incorporates network monitoring functions as well as remote network management facilities.

The overall structure of SMT and its relation to the established network management principles is discussed in this chapter. As shown in figure 7.1 the components of SMT can be divided into two parts:

• Local management. This is discussed in this chapter.

• Remote management. This is discussed in chapter 8.

7.2 PRINCIPLES OF NETWORK MANAGEMENT

Network management has gained increasing importance as networks have become increasingly heterogeneous, wide-spread, and sophisticated. In today's corporate environment, most functions are carried out via network computing, and corporations are spending significant sums of money to provide high-capacity fault-tolerant computers and networks. The management information systems (MIS) departments have to deal with a variety of disparate equipment, networks, protocols, and applications. This has accentuated the need for network management and led to network management becoming the buzz-word of the decade.

Network management is an aid to the MIS managers and system administrators in maintaining better control of the network, and monitoring and diagnosing problems on a network. In order to provide a uniform capability, the ISO network management standards have extended the basic OSI reference model to include a management framework. The OSI network management paradigm can be divided into five categories, known as specific management functional Areas (SMFA):

Fault Management. This provides the ability to detect, isolate, and (if possible) correct network problems.

Configuration Management. The enables network managers to monitor, change, collect data from, and provide data to managed objects for the purpose of providing a seamless and continuous operation of network services. This is a useful feature in an office environment where network equipment is widely dispersed.

Performance Management. This provides the facilities to monitor and evaluate the performance of the network. This includes measuring statistics such as network traffic, throughput, and delay characteristics at various points in the network.

Accounting Management. This provides control features such as billing users for network resources used and to limit the use of those resources.

Security Management. This concerns managing access control to the network, authentication mechanisms, encryption of data, and so on.

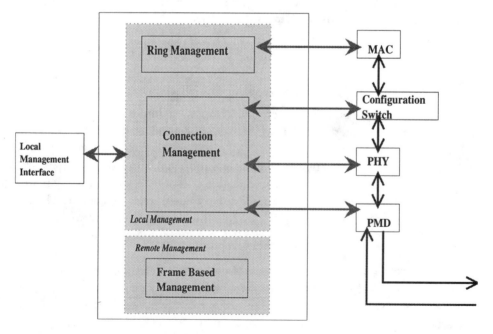

Figure 7.1 Components of station management.

Station Management Functions	Equivalent SMT State Machines
Network Level Functions • Protocols for remote management of stations • Fault isolation and recovery • Duplicate detection and resolution • Statistics gathering	Frame-Based Management (FBMT)
Node Level Functions • Station insertion (removal) • Station initialization • Configuration management • Fault isolation and recovery • Statistics gathering	Entity Coordination Management (ECM) Configuration Management (CFM) Ring Management (RMT)
Link Level Functions • Connection management • Fault isolation and recovery • Connection/resource availability policies • Statistics gathering • Noise and error monitoring	Physical Connection Management (PCM)

Figure 7.2   Functions and components of station management.

Of these five areas, FDDI explicitly provides fault management, configuration management (of the FDDI subnetwork), performance management, and interfaces for security management.

Very few networks today provide for explicit network management. They do provide for network monitoring. FDDI is unique in its ability to provide sophisticated remote management of FDDI stations.

7.3 FUNCTIONS OF SMT

Ethernet and Token-Ring evolved with essentially no monitoring or management capabilities. These capabilities were not seen to be essential then. As network sizes increased and multi-vendor environments became more common, the need for a common feature set became more apparent. In other words, it has become increasingly common to have systems platforms from one set of vendors and network connections from a different set of vendors in an interoperable environment. As a consequence of the mix-and-match approach, network management has become increasingly complex with little or no management tools.

FDDI is the first LAN to recognize this problem and address it. The result is the station management specification which provides the feature set to do fault isolation and recovery, noise and error monitoring, statistics, station insertion and removal (from the ring), local and remote node management, configuration management, and connection management, all in a standard fashion.

The station management specification can be organized into four parts as shown in figure 7.2:

- Link level management
- Node level management
  (Link and node level management together comprise connection management, as shown in figure 7.3.)
- Network level management
- Management information base

7.3.1   Station Insertion into a FDDI Ring

How is a station inserted onto a FDDI network? What is the sequence of events leading to an *operational ring*?

The flow-chart of an FDDI station inserting into an FDDI ring is shown in figure 7.4.

FDDI is a point-to-point series of connections at the physical level. This forms a daisy-chain of links with the signalling sequence as shown in the figures 7.5a to 7.5e. Consider an FDDI network to be the *sum of the parts*. Thus, if we start constructing the network, the order of creation would be:

- A link (fiber or copper) connection would be formed. This is *link connection* and is controlled by *link management*.

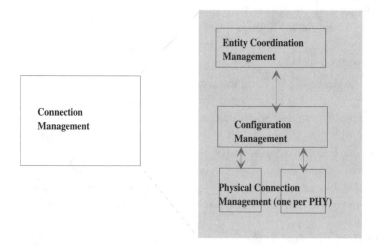

Figure 7.3   Components of connection management.

• Two nodes would then attach at each end of the link. This is *node connection* and it would have to be verified that the two nodes can communicate with each other. This is done by *node management*.

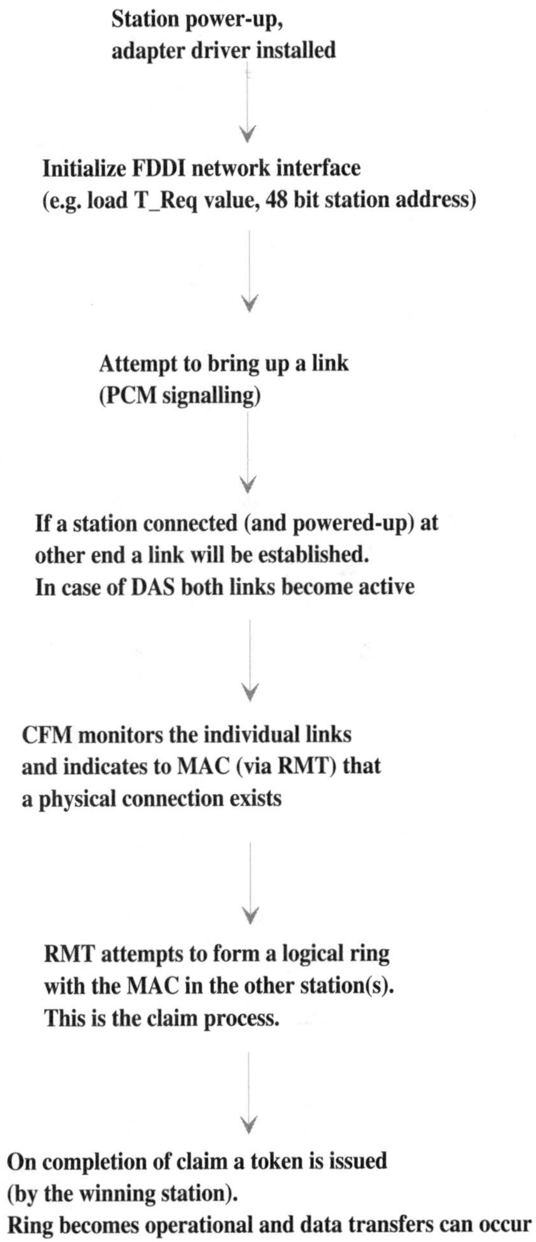

**Station power-up,
adapter driver installed**

↓

**Initialize FDDI network interface
(e.g. load T_Req value, 48 bit station address)**

↓

**Attempt to bring up a link
(PCM signalling)**

↓

**If a station connected (and powered-up) at
other end a link will be established.
In case of DAS both links become active**

↓

**CFM monitors the individual links
and indicates to MAC (via RMT) that
a physical connection exists**

↓

**RMT attempts to form a logical ring
with the MAC in the other station(s).
This is the claim process.**

↓

**On completion of claim a token is issued
(by the winning station).
Ring becomes operational and data transfers can occur**

Figure 7.4   Station insertion flow-chart.

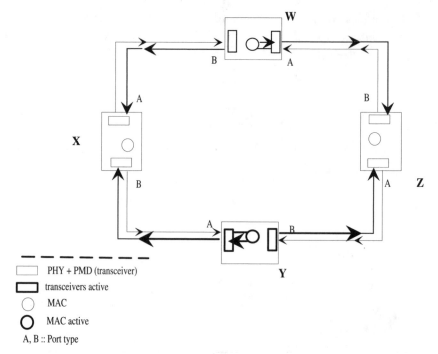

Figure 7.5a Station W and Y are powered on and attempting to connect. X and Z's network connections are switched off and are not attempting to connect.

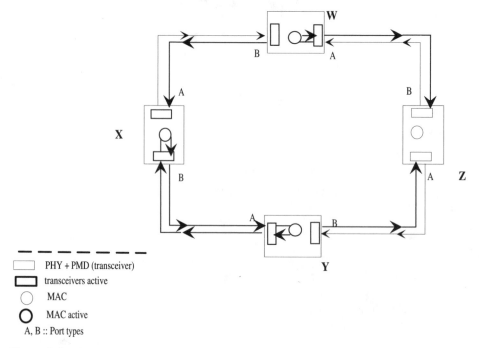

Figure 7.5b Station X's network connection is switched on and its B port link comes up first. Thus, the X-Y link is the first active loop.

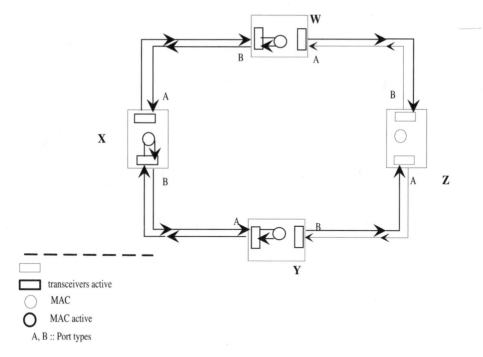

Figure 7.5c   A moment later, the A port of station X becomes active and the ring expands to W-X-Y. At this point, station Z is not yet active.

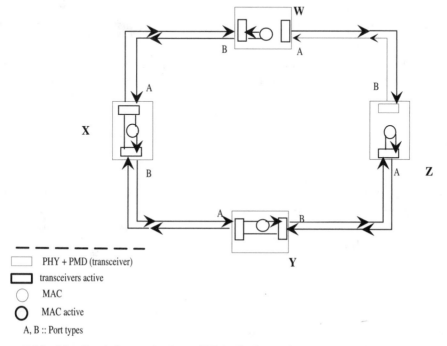

Figure 7.5d   After Z switches on, its A port link is the first to become active. The ring extends to W-X-Y-Z. At this time, the Z-W link is not yet active.

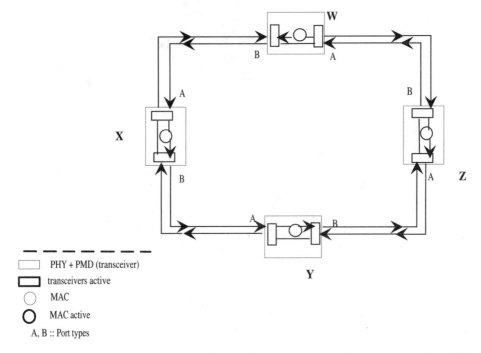

PHY + PMD (transceiver)

transceivers active

MAC

MAC active

A, B :: Port types

Figure 7.5e   A moment later, the A port of station X becomes active and the ring expands to W-X-Y. At this point, station Z is not yet active.

- This could be expanded by adding another link and another station (on either side of the two stations) until no more stations remain to be connected. This is controlled by *link and node management*.

- Once the ring is complete, it would have to be verified that every node can communicate with every other node. This is controlled by *network management*.

7.4   LINK LEVEL MANAGEMENT

Link level management ensures that a low bit error rate data-pipe is available for communication between two adjacent nodes. It ensures that the correct ports are connected together (to avoid redundant loops and wraps), and monitors the link for error-free operation. This process of building the network is performed by the following three state machines:

- Physical Connection Management (PCM)
- ConFiguration Management (CFM)
- Link Error Monitoring (LEM)

7.4.1   Physical Connection Management (PCM)

PCM is a state machine which controls the link connection. It ensures that a meaningful connection is established between two ports. In a dual attachment station there are two instances of PCM (one per port) because there are two sets of links, one attaching to the upstream neighbor and the other attaching to the downstream neighbor. In a single attachment station, there is only one instance of PCM because there is only one port and one link.

When an FDDI adapter, FDDI bridge port, or a concentrator is powered-on, the PCM state machine is enabled. The PCM state machine actively seeks to form a connection. In order for the PCM state machine to form a connection, two conditions must be satisfied:

- A physical link must exist between this station and an adjacent station.

- The station at the other end of the link must be powered-on and its PCM enabled.

If both conditions are satisfied, a connection is established and the port controlled by the PCM is said to be "up". Thereafter, the PCM is constantly executing and monitoring the link for errors or connection breaks.

7.4.1.1   The Need for PCM.   PCM is an integral and mandatory part of the FDDI network. PCM consists of the PCM state machine and the PCM pseudo-code. The state machine controls the synchronization of the connection, and the pseudo-code controls the information exchanged during connection initialization. PCM performs the following functions:

- By identifying the port types, PCM rejects unsuitable connections leading to undesirable topologies. Certain FDDI configurations are illegal because they create undesirable topologies such as more than two rings (see figure 3.4).

- Synchronization of the two ends of the connection.

- Performing a link confidence test before establishing a connection to isolate faulty links before the network becomes operational and ensuring that the network operates on links with a minimum BER.

- Establishing the identity of the nodes being connected. This may be accomplished by a mechanism called *MAC for local loop* and exchanging tokens and neighbor information frames before putting the port on the ring. This is useful in some applications where only certain stations are allowed to enter the ring.

Is PCM Really Needed?   At one time in the development of the FDDI network standards, an alternative type of configuration was suggested for low-cost, desktop connections. This configuration was called *Class C* wherein there was no PCM. Each station was a SAS with the receive coming from one node (upstream) and the transmit going to a different node (downstream). The connections formed a single uni-directional ring. This reduces the cost of an FDDI connection (desktop plus concentrator port) because it does not require a concentrator port for every desktop connection. This topology has seen some use in niche applications but has not become widespread because it compromises the fault-tolerance and error-detection capabilities of a node.

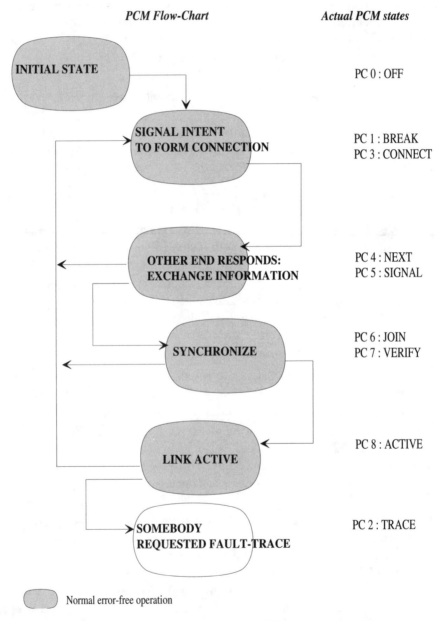

Figure 7.6   Simplified physical connection management (PCM) state machine.

7.4.1.2 The PCM State Machine. The PCM state machine has ten states. Part of this state machine is implemented in VLSI as it is time-critical. On power-up, the state machine is always initialized into the OFF state. A simplified PCM flow-chart with the corresponding states is shown in figure 7.6.

When it is desired to insert the node into the network, the adapter driver issues a connect signal, via SMT, which initializes the various components of CMT. This in turn triggers the PCM state machine (PC_Start is issued) of the ports within the node to establish a connection. Thereupon the PCM state machine transitions from its initial *off* state to *break* state.

The break state is essentially a state where zero or minimal light (quiet line state) is transmitted. This state is the first in a series of steps to a synchronized connection. In the break state, a continuous stream of quiet symbols is transmitted notifying the other end of the connection to stop transmitting data (if any is being transmitted) and enter break state. A transition out of break state is made after sufficient time for the node at the other end of the link to react to the Quiet symbols (TB_Min time-out). It transitions into *connect* state on receiving QLS or HLS from the other end of the connection.

The connect state is the first point in the sequence of states which lead to a synchronized active PCM connection. A port in the connect state transmits halt line state (HLS) continuously until it receives HLS from the node at the other end of the connection. Reception of HLS when transmitting HLS continuously indicates that the node at the other end of the connection is attempting to form a connection. connect state is thus an equilibrium state for a port which is willing and able to forge a connection but the node at the other end of the connection is either unwilling (node not initialized) or unable (problems) to do so. In order to avoid synchronization problems, on entry into connect state, the port transmits HLS for a minimum time-out (C_Min). This ensures that the node at the other end of the connection receives the signal (HLS). Receiving HLS from the other end of the connection when in connect state is the affirmative signal that both ports are synchronized and ready to establish a connection. Both ends of the connection will transition to the *next* state upon receiving HLS.

Micro-synchronization is performed in the next state. Whereas the connect state signalled the intent of the two ports to form a connection, the next state is where the actual mechanics of synchronization are performed. During this phase, important information is exchanged. This information is called the pseudo-code. The next state is like the pauses between words in a conversation. The pauses are idle line state (ILS). A port will alternate between pauses (next state) and words (*signal* state) until all the information of the pseudo-code is signalled. At the completion of the pseudo-code, if everything is all right, a flag (PC_Join) will be set, which causes a transition to *join* state.

The signal state signals bits to the node at the other end of the connection. The truth-table for the bits is:

0 implies master line dtate (MLS).

1 implies halt line state (HLS).

The signal state is entered from the next state whenever a bit of information is ready to be signalled. Every bit is signalled by a burst of MLS or HLS of TL_Min duration. Each bit of the information is punctuated by ILS transmitted from next state. Thus, an alternating stream of zeroes (MLS) and ones (HLS) and pauses (ILS) will be seen during the pseudo-code signalling. After all the bits have been signalled (normally ten), if there are no hitches, a flag (PC_Join) is issued indicating that the state machine should transition from next to join state.

Join state is the first of three states where various line states are transmitted to test the synchronicity of the connection. In this state, a continuous stream of HLS, followed by MLS and finally ILS is transmitted. On reception of HLS in join after transmitting for a minimum of TL_Min duration, a transition is made to *verify* state.

In the verify state, MLS (alternating halt and quiet symbols) is transmitted until MLS is received. At this point, the connection is complete, a transition is made to the *active* state and the port is ready to be incorporated into the token path.

On entry to active state, a continuous stream of ILS is transmitted. As soon as ILS is received in active (once again after having transmitted for a minimum time of TL_Min), the port is ready to be incorporated into the token path and issues a signal called CF_Join to CFM. CFM then proceeds to place the port on the path desired (primary, secondary or local). In the active state, only two line states are signalled: ILS and ALS. ALS is the active line state which is transmission of a frame. ILS is used to signal preamble or inter-frame gap (IFG). Thus, when the ring is operational (MAC viewpoint) or the link is active (PCM point of view), only frames and inter-frame gaps consisting of Idle symbols should be seen. Any other symbol or line-state during this state is an error.

A transition is made back to the break state if this node is initialized (PC_Start signal), if there is a break in the connection (QLS received), if the other end is initializing (QLS received), or if the PCM state machine times out while waiting for a transition in any other state.

Most transitions in the PCM state machine are bounded by an upper-bound of T_Out (default 100 ms). If a transition is expected and none is taken within T_Out (because the requisite line state is not received), then the state machine transitions to break state and the connection may be restarted. An example is a port stuck in verify state. In verify, the connection is undergoing its final synchronization and it should receive MLS within a very short time. However, if it is stuck transmitting MLS and no MLS is received for an extended duration, there is a problem in the connection and the connection is broken.

SMT also defines several optional transitions for enhanced error detection and recovery. These transitions basically allow for faster detection of problems.

7.4.1.3   PCM Pseudo-Code.   During connection initialization between the next and signal states, information is exchanged in the form of bit signalling. A total of ten bits of information is exchanged. Each bit is delineated by ILS and a one is HLS and a zero is MLS. The information exchanged is shown in the table 7.1.

Table 7.1   PCM Pseudo-Code

Transmitted Bit No.	Transmitted Bit Values	Description
0	0	This is the default value of this bit for this version of the standard.
	1	Transmitting a value of 1 for T_Val(0) is incorrect. However, a proprietary protocol may be executed when this bit is a 1. On receiving a 1, a node is allowed to discontinue signalling if its T_Val(0) = 0.
1,2	00	My port type is A;
	01	My port type is B;
	10	My port type is S;
	11	My port type is M;
3	0	This is the port compatibility bit and is not set if this node does not allow a connection to the port type signalled by the other end. (e.g. M to M connection);
	1	T_Val(3) is set if this node's topology connection rules allow connecting to the port type signalled by the other end. (e.g. A to B connection).
4,5	00	This signals a link confidence test (LCT) of short duration (50 ms default). On power-up, the initial test is of short duration.

Transmitted values are called T_Val()
Received values are called R_Val()

Table 7.1   **PCM Pseudo-Code (Continued)**

Transmitted Bit No.	Transmitted Bit Values	Description
	01	Medium duration (500 ms) LCT. This is signalled if the link has a history of LCT failures.
	10	Long duration (5 sec) LCT. This is signalled if the link has recently failed LCT or LEM (i.e. the LEM_Fail flag is true)
	11	Extended duration (50 sec) LCT. This is signalled if the link has a history of LCT failures and has recently failed LCT or LEM. This implies the link has persistent noise problems and essentially the connection is withheld (not allowed to become available for data transmission).
6	0	If reset, this bit indicates that no MAC will be placed at this end of the connection during the LCT and if a MAC is signalled to be placed at the other end, then this node will loop back the PDUs. If T_Val(6) = 0 at both ends of the connection, then the data sourced shall consist of ILS.
	1	If set, then this station shall place a MAC on this end of the connection. This implies that if both ends of the connection have a MAC available for a LCT then actual frames and token may be transmitted. If only one end has a MAC then the other end shall loop back the PDUs received.
7	0	LCT passed at this end of the connection.
	1	LCT failed at this end of the connection. Even if one end of the connection signals LCT failed then the connection is restarted and a long LCT is performed because of the recentness of the LCT failure.
8	0	A MAC is not available for local loopback at this end of the connection.

Transmitted values are called T_Val()
Received values are called R_Val()

Table 7.1   **PCM Pseudo-Code (Continued)**

Transmitted Bit No.	Transmitted Bit Values	Description
	1	A MAC available for local loopback at this end of the connection. During loopback the MAC transmission, recovery, and token passing processes are verified. Even the neighbor notification process is executed. This is typically used for graceful insertion or verification of the link and station at the other end before inserting into the ring.
9	0	If reset, it indicates that no MAC output will be connected to this port. This is useful information along with the PC_Mode (tree or peer) to trace a physical ring map of the internals of complex stations such as concentrators or DAS.
	1	If set, it indicates that a MAC output will be connected to this port. This can provide the same information as above.

Transmitted values are called T_Val()
Received values are called R_Val()

7.4.2 Errors, Link Confidence Test (LCT), Link Error Monitoring (LEM)

Errors in a network can be very troublesome. They can lead to retransmissions at higher layers and complicated upper-layer protocols in order to ensure error-free operation. These errors can be introduced by various sources such as connector and cable induced attenuation, dispersion, random jitter, duty cycle distortion, data dependent jitter (DDJ), interchannel isolation and crosstalk, and internal node noise. These errors can lead to corruption of data (loss of frames) and loss of bandwidth. If a large number of these errors occur, normal network operation can be hindered and can result in stoppage of data transfers.

In other LANs such as Ethernet and Token-Ring, it is extremely difficult to detect and isolate these errors without utilizing expensive equipment such as time-domain reflectometers (TDR) and error monitors. FDDI, however, has excellent error detection capabilities. These capabilities have been distributed in the PMD, PHY, MAC, and SMT specifications.

The PMD detects the loss of signal. The PHY detects line-state violations and clock rate differentials (elasticity buffer errors). The MAC has error detection capabilities such as frame check sequence (FCS), error indicator, lost counter, and error counter. The SMT monitors the PHY and interprets the presence or absence of symbols and line-states.

7.4.2.1 Types of Errors. In a FDDI symbol stream, the errors can be classified as:

a. Data symbol converted to data symbol (DD); a *1* corrupted to a *5*.

b. Data symbol converted to non-data symbol (DN); a *1* corrupted to a *J*.

c. Non-data symbol converted to data symbol (ND); a *K* corrupted to an *8*.

d. Non-data symbol converted to non-data symbol (NN); an *S* corrupted to a *K*.

The FDDI MAC is capable of detecting DD and DN type errors (and to some extent ND and NN errors such as E, A, C control symbols' errors). Since all FDDI nodes are not required to have a MAC, it was decided to require the PHY devices to be able to detect as many errors as possible. The FDDI PHY can detect DN, ND, and NN errors. For the proper operation of the FDDI network, it is necessary that all types of errors be detected. All types of FDDI nodes, including a concentrator with no MACs are capable of detecting and isolating most of the errors to the specific link on which they occurred (figures 7.7a to 7.7e).

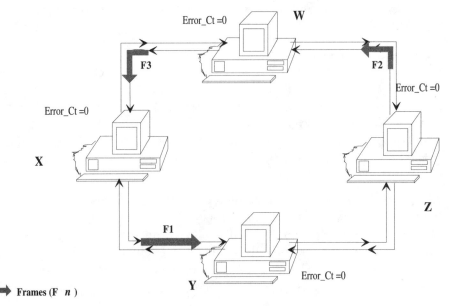

Figure 7.7a Initially, no errors on the network. All error counters are reset to zero. Normal ring operation is in progress.

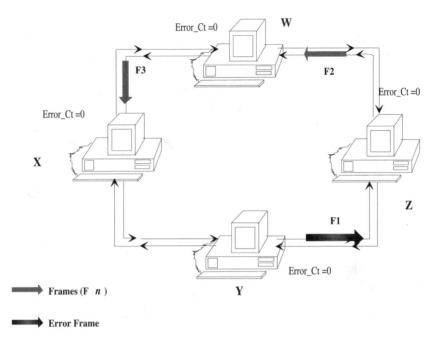

Figure 7.7b   Frame 1 is hit by an error after being repeated by station Y on link Y-Z.

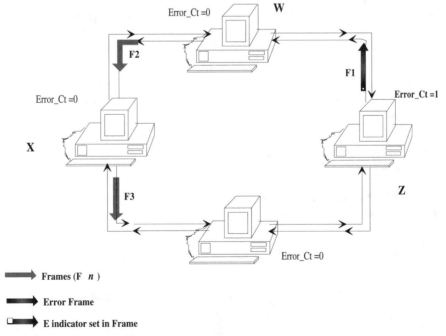

Figure 7.7c   The error on F1 is detected by station Z's MAC. It sets the E- indicator in the frame status field of the frame and increments its Error_Ct (to one.) Then it repeats the frame.

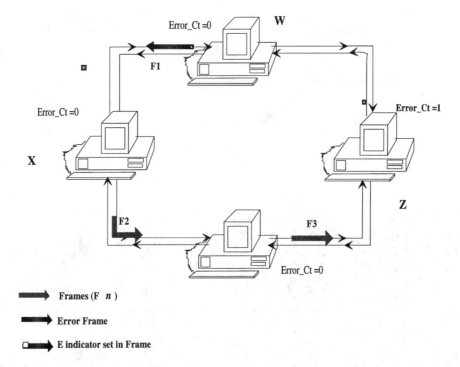

Error_Ct =0   **W**

F1

Error_Ct =0

X

Error_Ct =1

Z

F2

F3

Error_Ct =0

➡ **Frames (F *n*)**

➡ **Error Frame**

▭➡ **E indicator set in Frame**

Figure 7.7d   The error on F1 is detected by station W's MAC. However, it does not increment its Error_Ct because the E- indicator in the frame status field of the frame is already set, indicating error detected by an upstream station. It therefore repeats the frame without copying (if address matched).

Even if the errors occur between two MACless repeaters, the repeaters would provide the error statistics. This is an excellent feature which was recognized by the IEEE 802.3 committee and incorporated in the hub management standard. The FDDI MACless concentrators, hubs or repeaters are thus inherently *intelligent* (or *smart*) and can provide useful statistical information without a MAC.

7.4.2.2   What Are Link Errors?   Link error events, as defined in the SMT specification, can be loosely classified as:

• **Error 1**, which includes:

a. Transitions from ILS to LSU with the duration of LSU exceeding two symbol times (80 ns).

When a connection is active, only frames (JK FC nnnnn...) and interframe gaps (IIII....) can be transmitted. This implies that there can be only two PCM states: active line state (ALS) and idle line state (ILS). Therefore, there can be two transitions:

ALS to ILS: Frame followed by IFG.

ILS to ALS: IFG followed by frame.

The transition from ILS to ALS should not take more than two symbols (JK) as ALS is recognized by the symbol pair JK. Hence if line state unknown (LSU) is entered for a duration longer than two symbols, it is apparent that it is not the start of a frame that is being detected and hence is an error.

b. Transitions from ILS to LSU, the duration of which does not exceed two symbol times. These transitions should not include those caused by a JK symbol pair, or by the actions of the elasticity buffer or the decode function in the same PHY. This transition can be implemented if an implementation is capable of transitioning to ALS without transitioning to the intermediate LSU. This is optional in the standard.

• **Error 2**, which includes:

a. Transitions from ALS to LSU with the duration of LSU exceeding eight symbol times (320 ns).

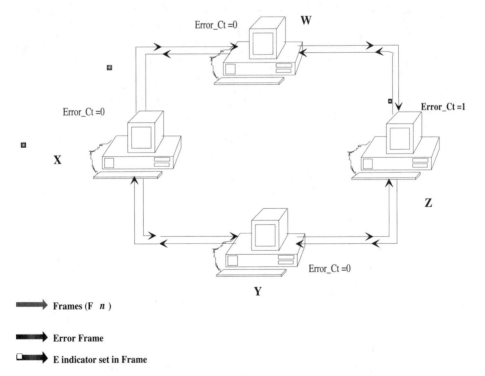

➡ **Frames (F _n_)**

➡ **Error Frame**

▭➡ **E indicator set in Frame**

Figure 7.7e   The frame thus propagates around the ring until it is stripped by the sender. At this time, if the Error_Ct in all the stations were to be examined by a network manager, it would be able to isolate the error to the Y to Z link because station Z's Error_Ct is one and Y's is zero.

When the PCM is in active state, a frame is normally followed by an inter-frame gap. Thus, from ALS (receiving frame) the PHY should transition to ILS (receiving inter-frame gap). But in order to recognize ILS, four bytes of idles need to be received. In the meantime, the PHY is allowed to transition to LSU. However, if the duration of LSU exceeds eight symbols, then an error can be indicated.

b.   Transitions from ALS to LSU with the duration of LSU not exceeding two symbol times.

These transitions also should not have been caused by a JK symbol pair (as in a token release immediately followed by a fragment (JK FC TTJKFCnnn...) or by the actions of the elasticity buffer or the decode function in the same PHY. This is also optional in the standard.

c.   Transitions from ALS to LSU provided that these transitions are not caused by a halt symbol, and the duration of LSU exceeds two symbol times. This is also optional in the standard.

7.4.2.3 Testing the Link at Initialization: Link Confidence Test (LCT).

The FDDI SMT specification provides for link error detection during connection initialization phase and continuous link monitoring thereafter. During the connection initialization phase, the PCM state machine includes a Link Confidence Test (LCT) which tests the physical link for errors. The LCT is conducted during the PCM pseudo-code signalling and can be accomplished by one of the following methods:

• Transmitting idle symbols and counting the number of link errors on the incoming link (figure 7.8).

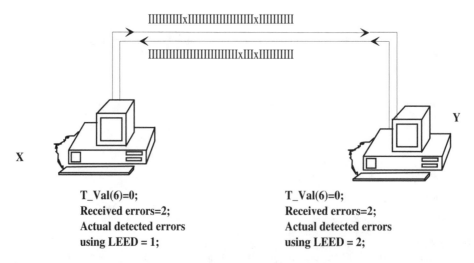

Figure 7.8 Transmit idles and count number of link errors on incoming link.

- Transmitting PDUs and counting link errors on the incoming link (figure 7.9).

- Transmitting PDUs and counting FCS errors (figure 7.10).

- Looping back symbols received from the other end of the connection and counting link errors on the incoming link (figure 7.10).

Method 1 is the minimum capability required of a node (and its ports). ILS has the maximum number of transitions of any symbol but it is not the best symbol sequence to detect data dependent jitter (DDJ) because it only contains one transition pattern (alternating zeroes and ones). This implies that this method will only detect a subset of the errors possible. This method does not require a MAC at either end of the connection.

In **Method 2**, PDUs are transmitted. This is better than Method 1 because data, control, and idle symbols are transmitted, alternating between ILS and ALS. However, the errors which are counted are the DN, ND, and NN types. The DD type of error can be detected by an FCS check only which is not performed in this case. Hence this method too detects a subset of the possible error-set. This method is employed when a MAC or some other data-source is placed on this end of the connection.

Method 3 detects the DD type of error. PDUs are transmitted and each incoming frame is checked for FCS errors. This check is performed at the MAC-level. This method has two drawbacks: most NN type of errors are undetected and multiple DN, ND errors are counted as single events. This method is employed when a MAC is available for the LCT.

Method 4 simply involves looping back the Phy_Data_Request (PDR) to the Phy_Data_Indicate (PDI) and counting link errors on the incoming link. The PDR is the MAC to PHY data path and the PDI is the PHY to MAC data path. This method is typically used when only one end of the connection can place a MAC on the path. The other end loops back the data stream received and simply checks the incoming data stream for link errors.

It is apparent that a combination of Methods 2 and 3 is the best method. However, that method is not implemented because it counts errors at two different levels (MAC and PHY) and it may count the same error more than once making it difficult to predict the actual error rate.

As we can see, LCT tests typically do not provide full coverage. As links degrade over time and there is a necessity to monitor the links for errors after connection initialization. This was recognized by the SMT committee and a post-connection initialization error monitoring capability was implemented. This is the link error monitoring function. This function shares the link error event detector with the LCT.

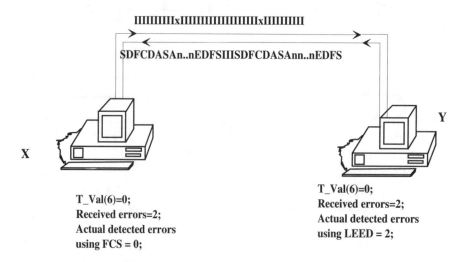

Figure 7.9   LCT- Y transmits frames and counts number of link errors on incoming link.

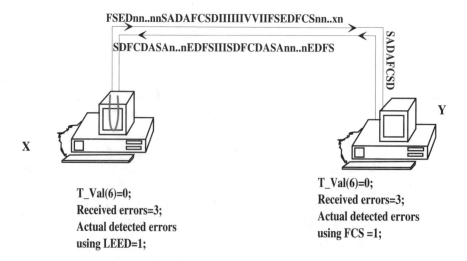

Figure 7.10   LCT- Y transmits frames and counts number of FCS errors on incoming link. X indicates corrupted data symbol and V indicates violation symbol.

7.4.2.4   Monitoring the Link When Ring is Operational: Link Error Monitoring (LEM).

The errors that can cause the most damage are the errors which occur when the ring is operational. During that time, there are only two possible line states: ALS and ILS. ALS is the frame duration and ILS is the interframe gap. Any other symbol or line state constitutes an error. To detect such errors, SMT requires every port to have a link error monitoring(LEM) function. Once the connection is active a link error monitoring function is implemented for continuous monitoring of the link. The LEM counts errors within a selected period of time (say T sec) and normalizes the errors to a rate function based on 125 Mbaud per second. Thus, the LEM outputs a link error rate (LER) based on the errors detected. The SMT document does not specify the interval over which the error rate should be calculated.

After the ring is operational, any state other than ALS or ILS is an error. If the LER exceeds a certain threshold (LER_Alarm), a signal is generated to the local management entity (LER_Alarm_Event). After the alarm, if further errors are detected and the LER_Cutoff threshold is exceeded, the link is declared faulty and the LEM_Fail flag is set. Typically, LER_Alarm is 10e-8 and LER_Cutoff is 10e-7. Table 7.2 illustrates the number of errors needed to exceed the various thresholds.

It can be seen from the table, even a few errors can cause the thresholds to be exceeded, thus causing the link to be rejected and connection withheld. A better method may be to add some hysteresis to the rate calculation such that the first few errors do not cause the link to be marked faulty.

7.4.2.5   Conclusions on LCT and LEM.

How does one quantify errors? The LCT and LEM error detection mechanisms are not capable of detecting all errors. A burst of errors may be detected as one error. Thus LCT and LEM provide a lower-bound on the actual error-rate. If JK FC nnn..TEACIIIIIIII JKFC nn... were to be converted to JK FC nnn..TEACI#I#I#I#IJK FC nn... , where # is violation symbols then LCT or LEM would only detect one error rather than four errors. The problem is even more fundamental–how does one calculate *actual bit error rates*? There is one school of thought which believes that line errors are bursty in nature and do not occur individually. There is another school of thought which believes that individual noise-hits are probable events. If one adopts the noise model which leads to bursty errors, it is possible to be more accurate in assessment of LERs. However, if one adopts a noise model which can lead to isolated errors, the LEM, LCT mechanisms can only be used to detect errors, not to specify an accurate BER. The LCT and LEM mechanisms are based on the model that noise is essentially bursty in nature. In all cases, LEM and LCT do not provide the actual BER.

Table 7.2   **Link Error Rates**

LER VALUE	N at T=1	N at T=.8	N at T=.5	N at T=.1	N at T=.005	N at T=.001
1.00E-15	0.000000125	0.0000001	6.25E-08	1.25E-08	6.25E-09	1.25E-10
5.00E-15	0.000000625	0.0000005	3.125E-07	6.25E-08	3.125E-08	6.25E-10
1.00E-14	0.00000125	0.000001	0.000000625	0.000000125	6.25E-08	1.25E-09
5.00E-14	0.00000625	0.000005	0.000003125	0.000000625	3.125E-07	6.25E-09
1.00E-13	0.0000125	0.00001	0.00000625	0.00000125	0.000000625	1.25E-08
5.00E-13	0.0000625	0.00005	0.00003125	0.00000625	0.000003125	6.25E-08
1.00E-12	0.000125	0.0001	0.0000625	0.0000125	0.00000625	0.000000125
5.00E-12	0.000625	0.0005	0.0003125	0.0000625	0.00003125	0.000000625
1.00E-11	0.00125	0.001	0.000625	0.000125	0.0000625	0.00000125
5.00E-11	0.00625	0.005	0.003125	0.000625	0.0003125	0.00000625
1.00E-10	0.0125	0.01	0.00625	0.00125	0.000625	0.0000125
5.00E-10	0.0625	0.05	0.03125	0.00625	0.003125	0.0000625
1.00E-09	0.125	0.1	0.0625	0.0125	0.00625	0.000125
5.00E-09	0.625	0.5	0.3125	0.0625	0.03125	0.000625
1.00E-08	1.25	1	0.625	0.125	0.0625	0.00125
5.00E-08	6.25	5	3.125	0.625	0.3125	0.00625
1.00E-07	12.5	10	6.25	1.25	0.625	0.0125
5.00E-07	62.5	50	31.25	6.25	3.125	0.0625
1.00E-06	125	100	62.5	12.5	6.25	0.125
5.00E-06	625	500	312.5	62.5	31.25	0.625
1.00E-05	1250	1000	625	125	62.5	1.25
5.00E-05	6250	5000	3125	625	312.5	6.25
1.00E-04	12500	10000	6250	1250	625	12.5
5.00E-04	62500	50000	31250	6250	3125	62.5
1.00E-03	125000	100000	62500	12500	6250	125

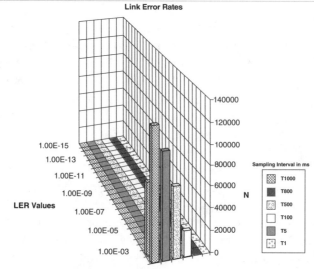

Link Error Rates

The time interval over which to calculate the actual LER has been a subject of debate but it is recommended that it not be too short or too long. Recommended values range from a few seconds to a few minutes.

7.5   NODE LEVEL MANAGEMENT

When at least one of the adjacent links is operational, node level management begins to function. There are two state machines defined for managing the node:

- Configuration management (CFM) or path management

- Entity coordination management (ECM)

At node level management, the functions provided are bypass switch control, placement of PHYs in the primary, secondary or local rings, control of the individual PCMs, and node level fault diagnostics.

7.5.1   Configuration Management (CFM)

CFM is the configuration management entity for the various resources such as Port or MAC. CFM controls the insertion and deinsertion of a port or MAC into the token-path. There is one CFM state machine per node MAC or PHY. The CFM state machine for each node is continuously and asynchronously executing. Any change in input conditions is evaluated by the CFM. In order to understand the functions of CFM, we need to understand the various FDDI topologies.

An FDDI network can have at most two logical rings or token-paths: primary and secondary. Primary denotes that the ring will be used for data transfer and the other ring (secondary) will be used as a backup. In a typical FDDI environment, data is circulating on one ring whereas idles are circulating on the second ring. Within an FDDI node, the two rings are called primary path and secondary path.

Since FDDI was originally formulated, a need for more flexible topologies was noted. For example, the US Navy has chosen to use both rings for data transfer in their version of FDDI (SAFENET). It is also feasible to have a single ring, hub-based, low-cost FDDI network. To accommodate the various configurations and to provide a more standard mechanism of accomplishing these configurations, the CFM state machines have been extensively reworked in SMT 7.2 and these configurations are defined on a per port or MAC basis. In a single MAC-SAS, only the S port and MAC CFM will be implemented. The A, B, or M port CFMs will not be implemented.

CFM performs the interconnections of the PHYs and MACs within a node where the node can be any of the following:

- Dual attachment station (DAS)

- Single attachment station (SAS)

- Dual attachment concentrator (DAC)

- Single attachment concentrator (SAC)

Requested path and *current path* are two terms defined in CFM. A resource can *request* to be placed on a path (primary, secondary or local). This is called the requested path. However, it may not be possible to place the resource on its *desired* path. A list of options can be provided which can be evaluated in a fixed order and the resource is placed on the first *available desired path*.

An S port CFM state machine is fairly trivial in a SAS. A SAS has only one token-path and it can either be *isolated* from the path or can be inserted (WRAP_S) on the path. Since a SAS has just one PHY (S port) and one MAC, the interconnection is straightforward.

A DAS has two PHYs (A and B ports) and one (or more) MAC, and the interconnection is more complex.

Similarly, the SAC and DAC entities can be pre-configured (or hard-wired) in a simple interconnection scheme or remotely manageable which provides for more flexible configurations. If the DAC supports multiple MACs or multiple rings, the CFM becomes extremely complex.

7.5.1.1   Paths.   A path is defined as a segment of an external ring which passes through a node. Although FDDI specifies two rings, a node may have more than two paths internally. The primary and secondary rings have corresponding internal paths which are no more than electrical traces. The additional paths are referred to as local paths and their control is not specified in the standard. A concentrator can support multiple rings using the local paths. The local paths are used for diagnostics and checking connections before connecting to the primary or secondary rings.

7.5.1.2   Remote Management of Paths and Entities on Paths.   The CFM controls the path and the switching elements on the paths. To enable sophisticated remote management, three types of management information base (MIB) attributes supporting paths are defined (see chapter 8):

- Requested path

- Available path

- Current path

Requested path is a per port (or MAC) attribute. The setting of the individual bits indicates the paths the port may be placed on. The bits are ordered so that if a path is unavailable the next bit which is set, is evaluated with the available paths attribute.

The available path attribute indicates the various paths (primary, decondary, local) supported in this node. This attribute does not change dynamically.

The *current path* attribute indicates the is the actual path of the port (or MAC) selected as a result of evaluating the requested path and available path attributes.

7.5.1.3 Configuration Switch. Every concentrator has a configuration switch which enables a concentrator to disconnect any or all of the attached nodes from the network. For multiple paths, the switch also allows the attached nodes to be placed on any of the paths and bypass any powered-off (or faulty) node. This switch is referred to as a configuration control element (CCE) in the SMT standard and is the definition of the interconnections (various paths) of ports (or MACs) to other ports (or MACs) within a node (figure 7.11). In the simple one-path SAS, a CCE is nothing but a metal trace connecting the PHY with the MAC with the PHY and/or the MAC both having the ability to bypass the path. In a complex, multi-port, multi-path concentrator a CCE is an actual multiplexer.

7.5.1.4 The CFM State Machines. Each resource (MAC or PHY) has an independent state machine which is waiting for any change in input conditions such as CF_Join, CF_Loop, Requested Paths, etc. These inputs can be triggered by the PCM state machine of a PHY transitioning to active or change in the MIB attribute requested paths.

The CFMs of each resource are not truly independent. An A port CFM cannot independently transition to the *thru* state. The thru state is reached by

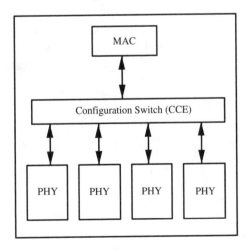

Figure 7.11   Concentrator configuration switch element or configuration switch.

the A port in lock-step with a B port. Thus, a DAS connecting to the trunk ring may see the CFM states for A and B ports sequence through ISOLATED → WRAP_A or WRAP_B → THRU. This information is usually available to the systems programmer through the debug interface of the FDDI device.

The A or B port CFM state machine contains {if...then...else} clauses which check if the port can be put in THRU, WRAP, C_WRAP, or ISOLATED state. It also checks if a connection needs to be withheld.

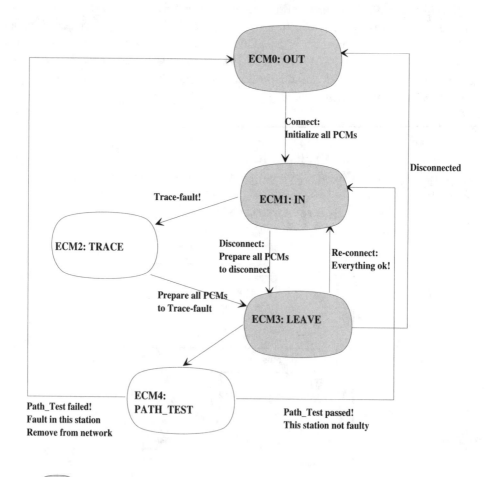

Figure 7.12 Simplified entity coordination management (ECM) machine.

7.5.2 Entity Coordination Management (ECM)

The reason for having a separate ECM state machine is tenuous at best. The ECM state machine does not provide much functionality other than acting as an interface to SMT requests to the individual PCMs and/or optical bypass. In the presence of a network_fault_trace condition, ECM performs a path test within a node and attempts to signal the trace upstream of the MAC. If the node passes the internal path test, ECM reinitializes the affected PCMs. In the presence of a node fault, or if a node is gracefully shutdown, ECM will activate the optical bypass switches if present. The optical bypasses are typically implemented in DAS or DAC configurations in order to maintain the dual rings in presence of node faults or node shutdowns. A simplified ECM state machine diagram without optical bypass control is shown in the figure 7.12.

7.6 NETWORK LEVEL MANAGEMENT

There are three components to network management:

- •Ring management (RMT), one per MAC
- •Management information base (MIB), one per node
- •Frame-based management (FBMT), one per station

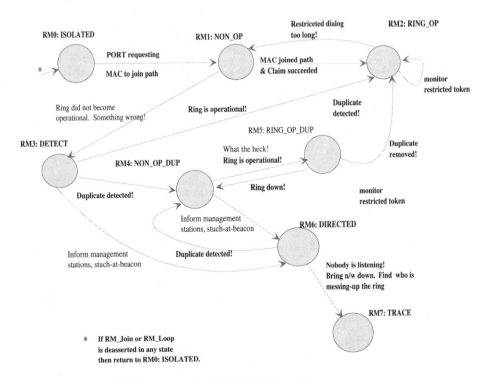

Figure 7.13 Simplified ring management (RMT) state machine.

CMT deals with the operation of the ports within a node. Network management deals with the operation of the nodes on a network and requires a MAC to operate. If a node has no MAC, *inband* remote management is not possible. *Inband* management is using the same FDDI network to perform remote management. *Out-of-band* management would for example use a RS-232 interface to the FDDI device to do remote management.

7.6.1 Ring Management (RMT)

RMT controls the placement and operation of a MAC on a ring. It is possible to have more than one MAC in a station. In a node with multiple MACs, there would be one RMT per MAC.

RMT provides MAC status information to SMT (or other management entities) and also detects and resolves *stuck-at-beacon*, and duplicate address problems.

A simplified RMT state machine is shown in figure 7.13.

7.6.1.1 Stuck-at-Beacon Fault. *Stuck-at-beacon* is the phenomena where a MAC keeps transmitting beacon frames. Normally, beaconing is a process which resolves on receiving ones own beacon frame. However, if a MAC does not receive its own beacon frame then it is stuck-at-beacon.

Stuck-at-beacon can only occur if there is a fault within the node on the receive path (or if the MAC chip is flawed). While the stuck-at-beacon problem may not manifest itself frequently, it is possible that this problem may arise in the field because of adverse operating conditions or prolonged exposure to extreme temperatures. The adapter may get heated up and/or the solder connection may come loose. In this situation, a MAC does not receive any valid data because of problems in its receive path. The TRT timer expires causing the MAC to start claiming. Claim will not resolve because no frames are being received. Claim will expire (T_Max expiration) and the MAC starts beaconing. After beaconing for some time, the RMT recognizes that the MAC is stuck at beacon and will start to transmit directed (to a multicast management address) beacon frames informing the manager that it is stuck.

Any other fault will not result in stuck-at-beacon at this node. If there is a cable fault, CMT will detect it. If there is a problem on the transmit path in a node, the next station downstream can get stuck-at-beacon.

The stuck-at-beacon problem normally resolves by one or more stations attempting to *trace* the network.

7.6.1.2 Trace Function. In trace, every station on the trace path performs a path_test which includes placing the MAC in a loopback path. If the station passes the path_test, it restarts the CMT state machines. If the station

fails the path_test, the station is isolated and the network configures around it (wraps).

The function of trace is to isolate the problem in the fault domain which is between two MACs on the same path. The trace function is best illustrated by the figures 7.14a through 7.14d. In the figures, station 7 in the tree is shown to be stuck-at-beacon. This causes the RMT timer T_Stuck = 8 ms to expire, causing the PHY to *initiate trace*. The PHY transmits master line state upstream of the MAC, indicating a problem with the station 7's MAC. This triggers the upstream PHY in station 6 to *propagate* or *terminate* the trace.

A trace is propagated if the PHY's input (PDR) is not connected to a MAC but to another PHY as in a concentrator. In propagating a trace, a PHY indicates to ECM that it received a *trace* which starts a Trace_Max timer and it requests the upstream PHY to transmit MLS to its upstream PHY. Since a MAC can get stuck-at-beacon only if its receive path or the upstream station's transmit path is broken, the fault is upstream of the MAC. Its domain extends up to the first upstream MAC that is encountered on the path.

A trace is terminated if the PHY's input is tied to a MAC output. In this case, the fault-domain is covered and trace is terminated at the MAC. It indicates trace termination by transmitting QLS on the PHY that received the MLS. Following that, the station undergoes a self-test where it checks all the internal data-paths. On successfully completing the self-test, it restarts the connection sequence.

Once the fault is isolated (station 7 in figure 7.14d fails its self-test), the rest of the stations reestablish a ring (albeit a wrapped one) around the faulty station (figure 7.14e).

It is possible for a faulty station to be unable to trace. This would cause some other station to time out and initiate trace which would eventually succeed in wrapping around the problem node.

7.6.1.3 Duplicate Address Problems.
Duplicate addresses in any network can cause havoc on the network. Not only is the integrity and security of the network compromised, it is very difficult to isolate the duplicate station(s). RMT has enhanced duplicate address detection and resolution techniques to alleviate problems before and after ring operational. Note that duplicate addresses can prevent the ring from becoming operational during the claim and beacon process.

Detecting duplicate addresses is not a guaranteed process. That is why it is possible for the ring to become operational in the presence of duplicates. Over time it is always possible to detect the problem. Duplicates can be detected in any number of ways:

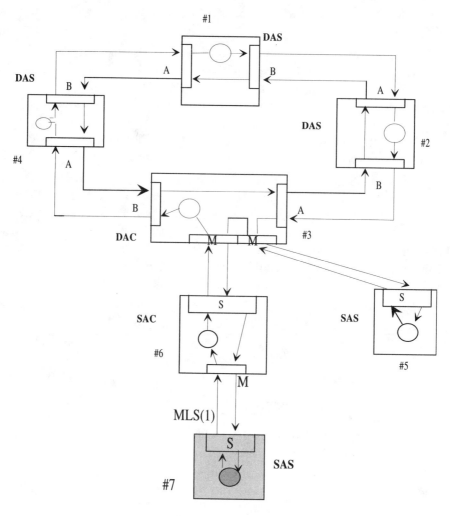

MAC "Stuck-at-Beacon"; station #7 initiates trace

Figure 7.14a Station 7 stuck-at-beacon and initiates trace.

- Receiving own claim frame long after (2 * D_MAX) this station stopped claiming. The longest time that it should take for a station to receive its own frame after transmission is the ring latency, which is D_Max. If the two rings have wrapped to produce one very large ring then the maximum time taken is 2 * D_Max. Thus, if own claim frame is received long after the station last issued claim, then it is a duplicate station issuing claim.

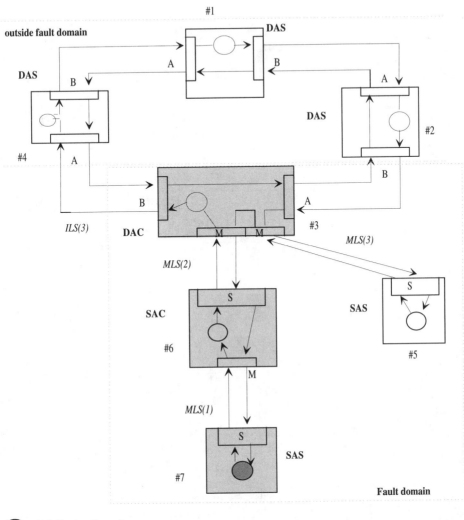

Figure 7.14b   Stations 6 and 3 propagate the trace because a MAC is not inputting to the port receiving trace.

- Receiving own beacon frame long after (2 × D_MAX) we stopped beaconing. The same reasoning as above applies for beacon frames too.

- Receiving own claim frame with different bid values (T_REQ).

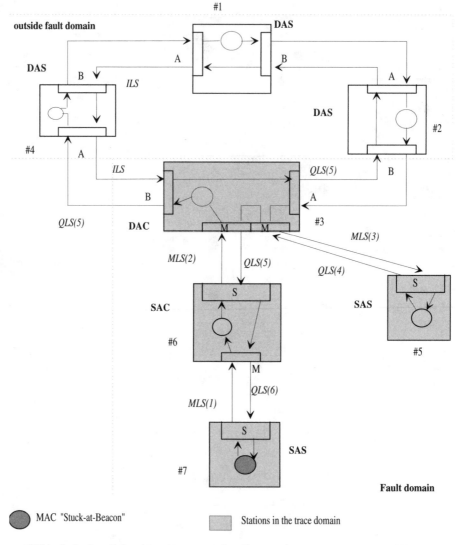

Figure 7.14c Station 5 terminates trace (MAC inputting to PHY) and signals QLS. Station 3 goes into self-test and propagates QLS on every out-going port on that path. Station 6 follows into self-test a little later, while stations 4 and 2 restart the connection.

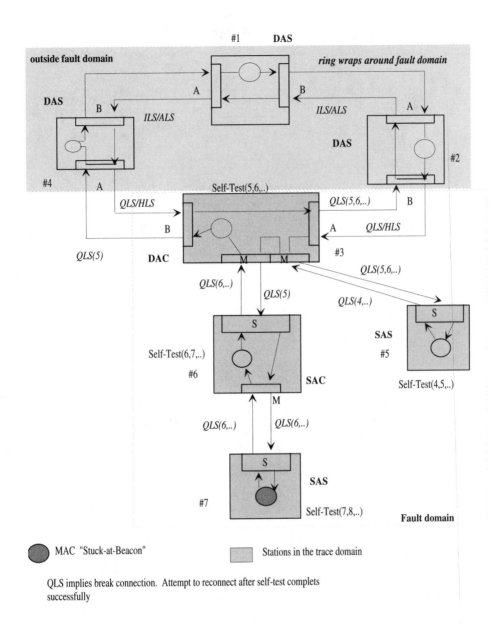

Figure 7.14d   Stations 3, 5, 6 and 7 are in self-test. Stations 1, 2 and 4 are not in the fault domain and wrap into a separate ring.

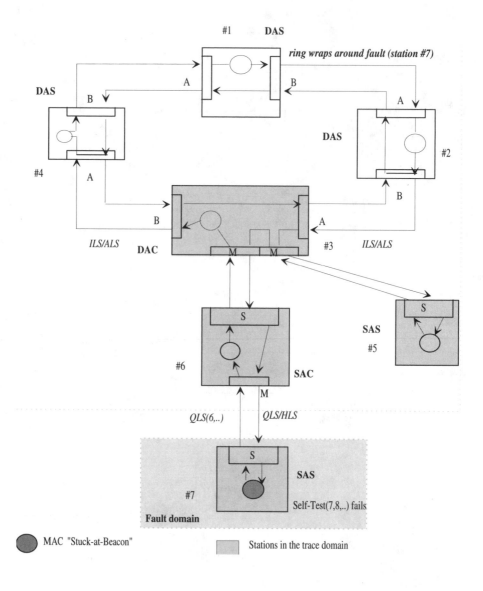

Figure 7.14e Stations 1, 2, 3, 4, 5, and 6 wrap around faulty station 7. Station 7 has failed the self-test and is automatically isolated from the network.

- When this station receives a claim frame with SA = My Source Address but the T_Bid! = T_REQ, then there is a duplicate station on the network.

- *Receiving an SMT NSA frame* (see chapter 8) *with DA = My SA and A indicator set.*

It is possible not to detect duplicates during claim and beacon process or to detect duplicates during pre-ring operational and proceed to ring operational. Once the ring is operational, SMT specifies a class of SMT frames which can be used for duplicate detection. In this mechanism, if a frame with a special Frame Control field (SMT_NSA = 0 × 4F) and DA = My SA is received from an upstream station with the A indicator set, implies that a duplicate upstream of this station copied the frame. What happens if the duplicate is downstream of this station?

- Receiving a frame with SA = My SA, DA = broadcast and A indicator not set.

This condition indicates that the upstream neighbor is a duplicate or there is no other station on the ring. There is one case where such a condition does not imply a duplicate: if it is the only station on the ring!

Once a duplicate is detected, FDDI ensures that network operations are unaffected for the other non-duplicate stations. If a duplicate is detected, the duplicate stations must disable LLC (or upper layer) services until the duplicate address problem is resolved. Disabling LLC services avoids problems with upper-layer protocols which might time-out because no response is forthcoming from the destination (which never received the frames because a duplicate stripped off the frames before they could reach the destination). Moreover, the TVX timers can expire in some or all of the downstream stations as they will not see any valid frames for extended periods of time. The problem can be resolved in a number of ways:

- *Change source address.* This requires a station to have more than one 48-bit (or 16-bit) address allocated. The station should also ensure that all PDUs sourced under the duplicate address are removed before changing the address. This mechanism is especially useful when the station is using 16-bit addresses and a duplicate is detected. However, multiple 48-bit addresses may be allocated to critical stations (e.g. file servers, routers, etc.) on the network.

- *Jam the network with a type of beacon frame (jam) while configuring the MAC to lose the claim process to other MACs.* This ensures that the ring is not prevented from becoming operational and the rest of the FDDI nodes from communicating. The MAC has to be configured to lose the claim process in order to avoid the duplicates issuing duplicate tokens or preventing ring operation. In the *jamming* process, a special type of beacon frame with Destination Address = My Long Address and the first octet of the info field containing the Beacon_Type = 0×02 followed by three octets of padding (zeroes). The jam beacon is sent for a sufficiently long time to ensure that the other duplicates take corrective action as well. There are two ways of jamming: One is disable the MAC input while sourcing beacons and the other is to perform jamming twice (thus ensuring that all duplicates will be notified). Both methods ensure that all duplicate MACs are notified and corrective action taken by all.

•*Remove the duplicate from the network.* Once the duplicates are removed from the network, the problem is resolved.

Of the three mechanisms, the third is the simplest to implement and probably the most effective. Most implementations execute the third alternative. Only stations which cannot be removed from the network (e.g. file servers, routers) will normally attempt the first or second methods.

The RMT state machine is currently not implemented in silicon but in software. After the SMT standard stabilizes, RMT may well be integrated into the FDDI silicon.

CHAPTER 8

Remote Management with FDDI

8.1 INTRODUCTION

LAN manageability is a very important requirement in the corporate network environment. In chapter seven, we discussed the features embedded in the station management specification of FDDI for local management at the node and link level. The MIS manager or system administrator is more concerned with his or her ability to remotely monitor and configure a network. This is remote management. There are two requirements for remote management:

•An information database.

The information database for each networking element such as bridge, router, adapter, and repeater has to be defined in a standard manner to provide interoperable and open management capabilities. FDDI defines a management information base (MIB) for FDDI devices and elements which is explained in this chapter.

•A frame-based protocol to obtain and change attributes.

A very popular frame-based protocol is the Simple Network Management protocol (SNMP) which is used extensively in management of Ethernet, Token-Ring and FDDI networks. In addition, FDDI defines a set of frames and frame-based protocols to perform the remote management. We discuss the FDDI-defined frame-based protocols, SNMP and ISO-based remote management protocol – Common Management Information protocol (CMIP).

8.2 MANAGEMENT INFORMATION BASE (MIB)

In order to provide extensive remote and local management of a node, an information database has to be defined. Like any other database, the information in a node is organized in a hierarchy (figure 8.1). The first level of the hierarchy are *objects*. Objects are *macro pieces of information*, for example, routing tables, MAC, PHY. These objects represent a collection of information. The smallest *atomic* piece of information contained within an object is called an *attribute*. A similar collection of attributes is called a *group*; for example, counter group: Error_Ct, Lost_Ct, Frame_Ct and so on.

Often, it is necessary to have a further classification: A bridge implements a *package* of attributes and/or attribute groups and an FDDI station implements another package of attributes. Both the bridge and station packages may share attributes, but each has a sufficient number of attributes within the group that are different and unique to a station or a bridge. Packages offer a simple and convenient method of clearly specifying the attributes necessary for each object when the attributes may be shared across different objects.

A hierarchical structuring of information allows for efficient retrieval and storage of information. This hierarchical database is called the management information base (MIB). The MIB has often been termed as a *view of the node*. Network management has become a very important issue in networking today and most networks specify a MIB. Since the original Ethernet specification did not include a MIB, the networking community defined an Ethernet MIB in the IETF committee. This standard definition specifies the object information that is contained in an Ethernet node.

8.3 STRUCTURE OF MANAGEMENT INFORMATION (SMI)

How is the MIB defined? What information can or cannot be in the MIB? How is the information used? These questions are answered in an OSI document called SMI. The SMI specifies how to classify and structure information, and defines a standard way of defining various types of managed objects, such as counters, timers, events, thresholds, and states. It uses an object-oriented information model to describe a managed object. The object-oriented information model defines the main concepts, terminologies, and structures. SMI defines a universal definitions of management information (UDMI) and

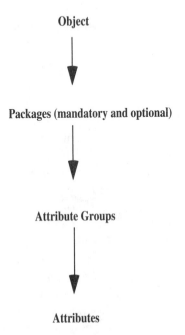

Figure 8.1 MIB organization.

generic definitions of management information (GDMI). The UDMI describes a number of types of objects and attributes that may be components of managed objects (e.g. MAC) for a wide variety of classes (FDDI, Ethernet, IP). For each object element and attribute element, the definition covers the structure, permitted operations, inherent properties, implicit relations with other information elements, and the specification properties of an element (state, counter, gauge, counter-threshold, event, etc.). The GDMI provides a template for defining management information specific to particular layers of applications. The essence of the GDMI is that it describes the properties common to potentially all objects or attributes of a class (e.g. MAC). The FDDI MAC would use the GDMI template for MAC to define an FDDI MAC object.

8.4   ABSTRACT SYNTAX NOTATION ONE (ASN.1)

The language used to describe the MIB information (such as objects, attributes, attribute groups, packages, etc.) in a machine independent format is the ASN.1. This is similar to using the C language to write *portable* programs. The ASN.1 is a programming language which defines complex data-types beyond bits and bytes such as integer, octet-string, bit-string, and so on. As with any other high-level computer programming language (HLL) ASN.1 also defines standard rules of coding. Among the most important and fundamental concepts of ASN.1 are type and value. A *value* is the information content of a particular data structure (e.g. integer with value 20). A *type* is a set or class of one or more values (e.g integers). ASN.1 also defines sub-types (e.g. integers 5 through 10).

ASN.1 has a number of built-in types (e.g. integer, octet-string). ASN.1 also defines a number of tools with which the user can define additional constructed types as well as sub-types.

As with any other language, different protocols use different subsets of the ASN.1. This is because the ASN.1 is *everything to everybody* which can be quite cumbersome. To make the protocol and the MIB definition simple, most protocols use smaller subsets. Thus, the Simple Network Management Protocol (SNMP) uses a small subset of the data-types and encoding rules. Similarly, FDDI uses a small but different subset of the data-types and encoding rules. This has the unfortunate side-effect that if the FDDI MIB definition uses a different data-type which is not used in SNMP then a translation is required between the SNMP and FDDI protocols attempting to access a shared database.

8.5   BASIC ENCODING RULES (BER)

Once the database is defined using ASN.1, a format needs to be specified on how it will be communicated across open systems (e.g. the network)–a *transfer syntax.* This includes specifying the byte to be transmitted first, the order of transmission of the byte, the type of data (octet, integer, bit-string, etc.), length of data (number of octets of information), and the actual value of the data. This is specified in the basic encoding rules, commonly referred to as BER. BER is then used to specify the network transmission format of the ASN.1 MIB information in frames.

BER does not provide a unique mapping. This makes it difficult for applications requiring a unique mapping to use the BER as is. Each protocol implements different flavors of BER and hence mappings have to be provided between the protocols. An example mapping issue is discussed later in section 8.8 where a station implements the ISO Common Management Information protocol (CMIP), the SNMP, and the FDDI SMT frame-based management protocols.

FDDI SMT encoding rules although similar to BER are different in many respects. The encoding of SMT TLV and BER TLV are different. In SMT, the type is tagged as a one byte field.

8.6 FDDI MIB

FDDI has SMT, MAC, PORT, and PATH objects (figure 8.2). The SMT managed object class models the management information for an FDDI station (a node with a MAC). There is only one SMT object per station and it contains attributes such as Station ID, MAC and PHY counts, and ECM state.

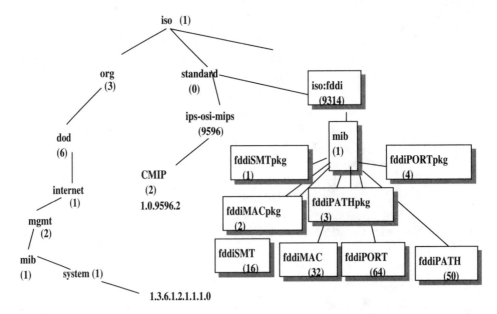

Figure 8.2 MIB hierarchy.

8.6.1   FDDI MAC Object

The MAC object class models the management information required for the MAC entities within a station. Since there can be multiple instances of the MAC object (up to 255), there will be multiple instances of the MAC object in the MIB. The MAC object class has attributes such as MAC address, Frame-Ct, Downstream Neighbor, and so on.

8.6.2   FDDI PORT Object

The PORT object class models the management information required for the PHY and PMD entities within a node. There can be multiple ports in a station (up to 255). A dual attachment station would have two instances of the PORT object in the MIB. Examples of PORT attributes are My Port Type, Neighbor Port Type, Port LER Estimate, Current Path, and so on

8.6.3   FDDI PATH Object

The PATH object class models the management information required for the PATHs within a node. A PATH is the *metallic bus* within a station which connects the PORTs and MACs together. In a SAS configuration there is only one PATH within a station. Concentrators can have one, two, three or more PATHs. A DAS or DAC will have at least two PATHs: primary and secondary. The primary and secondary paths are manageable in the FDDI framework. Additional paths, called local paths, are not defined in the SMT standard and are not manageable by standard SMT protocols.

8.6.4   Types of Attributes

The FDDI MIB consists of the following information types:
- Attribute
- Attribute groups
- Special attributes such as actions and notifications

FDDI attributes follow a naming convention of *Network.MIB object. attribute_name.* for example, fddiSMTStationId or fddiMACError-Ct.

Moreover, in the case of a station with multiple MACs or PORTs, multiple *sets* of attributes exist for each *instance* of the MAC or PORT. How does one distinguish between the multiple instances of the object? FDDI has a name binding scheme wherein an index is used to distinguish different MACs or PORTs. In order to access the fddiMACError-Ct attribute of the second MAC in the station, the MACIndex (= 2) attribute would be used.

FDDI SMT defines about 130 attributes in the MIB. These attributes are mandatory (e.g. fddiSMTStationId), optional (e.g. fddiMACT-Pri*n*), or mandatory if the underlying hardware is present (e.g. fddiPORTConnectionPolicies).

Each of these attributes have different access rights. Some attributes might be read-only (GET) and others might be read-write (GET-REPLACE). A read-only attribute (GET) can be fetched remotely over the network or over the local interface (within the station) but it cannot be changed. A read-write attribute (GET-REPLACE) can be fetched remotely or locally and can be changed remotely (if a station has the right to do so).

To allow vendors to offer enhancements and product differentiation, the standard allows private attributes to be defined within the FDDI framework. Because of the *private* nature of the definition, these attributes cannot be read by all implementation. The vendors may publish these attributes and their encodings or implement a MIB parser which can parse the vendor specific attributes. This is similar to the SNMP and Ethernet MIB-II paradigm. It is important to note that these private attributes cannot duplicate any standard MIB attribute.

Certain attributes are grouped together so that they can be accessed using a single command. These are called attribute groups. Attribute groups allow a remote manager to access n attributes in one frame rather than make n requests for n attributes and receive n responses. This has a two-fold advantage: reduced network traffic and ease of management. Most of the attributes in the FDDI MIB belong to some attribute group such as fddiMACOperation-Grp, fddiSMTStationIDGrp, and fddiPORTLerGrp. The only attributes that do not belong to a group are the vendor specific attributes, actions and notifications. There are 18 attribute groups defined in the four FDDI objects. A detailed description can be found in the ANSI SMT document.

There are certain special attributes called *actions* and *notifications*. Actions can be thought of as an emergency management mandate. Actions typically cause a station to disconnect, reset, or disable the MAC to LLC interface. For example, the fddiPORTAction can cause a PORT to start or stop. There can be two types of actions defined per object: A standard action and a vendor-specific action. The vendor-specific action allows for product differentiation and enhancements. Vendor-specific actions cannot duplicate the standard action.

8.7 FRAME BASED MANAGEMENT (FBMT)

Management can be performed locally or remotely. In local management, a network or system administrator performs the management functions at the managed node itself. The administrator can manually change certain attributes, change configurations, and start or stop processes via the keyboard or a utility (script or shell program). Local management is feasible in a small environment where the number of network entities is less than 20 or 30. In any environment with a large number of nodes or multiple protocols (e.g. TCP/

IP, Netware, etc.), it becomes significantly more difficult to manage each node on a local basis. In these instances, remote management is used.

In order to obviate problems with multi-protocol network management, FDDI SMT specifies a subnetwork-specific frame-based management protocol suite. All stations on an FDDI network can be managed remotely by the use of SMT frames. The transmission and reception of the SMT management PDUs does not require the use of any upper-layer protocols. This allows for uniform management of an FDDI network.

8.7.1 SMT PDU Contents

The SMT PDU content is as shown in figure 8.3. The SMT PDU is encapsulated in a FDDI MAC header (start delimiter, frame control, destination and source address) and trailer (frame check sequence, ending delimiter and frame status indicators). The SMT PDU can be distinguished by the value in the FC field.

8.7.2 SMT Frame Control Field

The frame control (FC) field distinguishes six major frame types:
- Token
- Void
- MAC
- SMT
- LLC
- Implementer

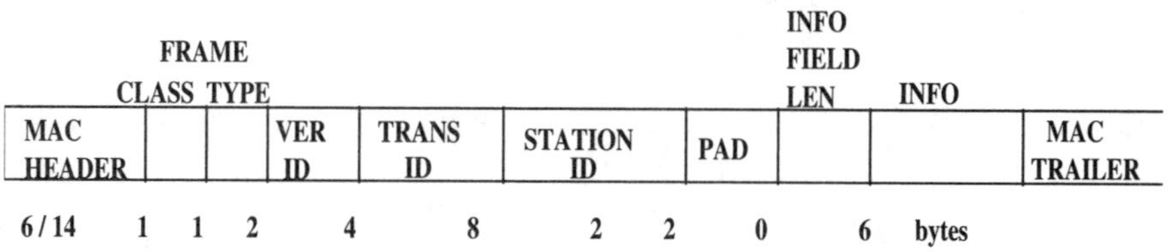

MAC HEADER			VER ID	TRANS ID	STATION ID	PAD	INFO FIELD LEN	INFO	MAC TRAILER
FRAME CLASS	TYPE								
6 / 14	1	1	2	4	8	2	2	0	6 bytes

· There are 12 frame classes defined.

· Frame types are Announce, Request and Response.

· Version ID is currently set to one.

· A unique ID number used for pairing Requests to Responses

· Station ID: *yy yy xx xx xx xx xx xx*

                    yy = implementer defined octet

                    xx = IEEE MSB order 48 bit address

                    Useful in domain-based addressing or complex FDDI stations.

Figure 8.3 SMT frame content.

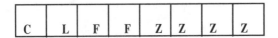

C = 0 indicates an asynchronous frame

C = 1 indicates a synchronous frame

L = 0 indicates 16 bit address

L = 1 indicates 48 bit address

F = Frame format bits; used in conjunction with CL bits
Z = Control bits; define individual frame types and priorities

SMT frames ::= 0100 0001 to 0100 1111; 0x4F and 0x40 used.
MAC frames ::= 1000 0001 to 1100 1111;
LLC frames ::- 0001 0000 to 1101 1111; 0x50 and 0x51 used.

Figure 8.4   Frame control field.

The SMT FC fields can take on any value between 0x01 to 0x4F. This is the block of FC values allocated by the MAC standard for station management purposes. The current SMT standard requires the use of long addresses and hence the FCs are limited to 0x41 to 0x4F (as shown in figure 8.4). Of these, only 0x41 and 0x4F are used by current SMT specified management protocols.

The FC = 0x41 is called the SMT_INFO FC and the FC = 0x4F is called the NSA FC (next station addressing). Several SMT protocols use the NSA FC. A typical NSA protocol would transmit a SMT frame with FC = NSA, DA = Broadcast, and SA = My Long Address. Every station on the network can potentially copy the frame but only the *immediate downstream station* can respond to this frame. This is because the first station downstream copies the frame (DA = Broadcast), *sets* the A and C bits or indicators, and repeats the frame. The following stations (may[1]) copy the frame (DA = Broadcast) but will not respond to it because the A indicator is received as set.

8.7.3 Addressing

The destination and source addresses in SMT frames use only 48-bit addresses and are similar to other frames (LLC, MAC, etc.). If the addresses are carried as attributes within the information field, the order of transmission is FDDI MAC order (msb of the MSB first) and not canonical.

---

1. Some silicon implementations have a mode for filtering of the NSA frames. The NSA frames are only copied if DA = My Address (which includes broadcast address) and A indicator = Reset. If both conditions are not met, the frame is not copied.

The only other requirement on SMT addresses is that the SMT response frames shall always contain an individual address, that is, responses shall be directed to the requestor which is always an individual station.

Several special addresses are defined for use within SMT protocols. The IEEE has set aside a block of 16 addresses in the block

0x 80:01:43:00:80:0y (in MSB form); where y = 0x0 − 0xF.

Of these, three addresses have been assigned so far (table 8.1).

Table 8.1   **SMT Multicase Addresses**

Canonical Address	MSB Representation	Assigned To
01-80-C2-00-01-00	80-01-43-00-80-00	SMT-DB-DA[a]
01-80-C2-00-01-10	80-01-43-00-80-01	SMT-SRF-DA[b]
01-80-C2-00-01-20	80-01-43-00-80-02	SMT-SBA-DA[c]
01-80-C2-00-01-30 to 01-80-C2-00-01-F0	80-01-43-00-80-03 to 80-01-43-00-80-0F	Not Assigned

a. SMT-DB-DA SMT directed beacon destination address
b. SMT-SRF-DA SMT status report frame destination address
c. SMT-SBA-DA SMT synchronous bandwidth allocation destination address

There is another address assigned to SMT. In every other network, 0x00-00-00-00-00-00 would be considered a *null address*. Not in FDDI! FDDI MAC allows the use of the null address under special circumstances for certain protocols such as duplicate address detection. The question then was how to represent *unknown addresses* (e.g. when the neighbors' address is not known). Instead of representing unknown by 0x00-00-00-00-00-00, a special 48-bit address was selected. This address was volunteered by DEC from their OUI space and is 0x00-00-F8-00-00-00 (in canonical form) and 0x00-00-1F-00-00-00 (in MSB form).

8.7.4   SMT Header

The SMT frame is encapsulated within the MAC header and trailer. It includes various fields to identify the frame category (frame class), the identification of request, response or announcement frame (frame type), the version of the SMT frame protocol (version ID), a unique frame identification field (transaction ID), the unique identifier of the station originating frame (station ID), a pad, and the actual parameterized information (figure 8.3).

8.7.4.1   Frame Class.   There are eight frame classes currently defined:[2]

1. Neighbor information frames (NIF)
2. Status information frames (SIF)
    • Configuration
    • Operation
3. Echo frames (ECF)
4. Resource allocation frames (RAF)
5. Request denied frames (RDF)
6. Status report frame (SRF)
7. Parameter management frames (PMF)
    • Get
    • Set
8. Extended service frames (ESF)

These frame classes can be an announcement, request, or response frame type. The combinations and permutations on these can lead to a lot of frames! As it turns out (thankfully) only a few frame classes and types are actually used by most stations. The rest are available for use by management stations or *sophisticated* stations.

Table 8.2   SMT Frames

Frame_Class	Encoded As	Frame_Type Encoded As	Required
NIF	0×01	A[a]	O
		Rq	M
		Re	M
Configuration SIFs	0×02	Rq	O
		Rs	M

Frame_Type: Required:
   A = Announce, 0×01 0 = Optional
   Rq = Request, 0×02 M = Mandatory
   Rs = Response, 0×03

---

2. Six frame classes have been added since SMT 5.1, the first landmark version of SMT. Most were added in SMT 6.2.

Table 8.2   **SMT Frames (Continued)**

Frame_Class	Encoded As	Frame_Type Encoded As	Required
Operation SIFs	0 × 03	Rq	O
		Rs	M
ECF	0 × 04	Rq	O
		Rs	M
RAF	0 × 05	A	O
		Rq	O
		Rs	O
RDF	0 × 06	Rs	M
SRF	0 × 07	A	M
Get PMF	0 × 08	Rq	O
		Rs	M
Set PMF	0 × 09	Rq	O
		Rs	O
ESF	0 × D	A	O
		Rq	O
		Rs	O

Frame_Type:
  A = Announce, 0 × 01
  Rq = Request, 0 × 02
  Rs = Response, 0 × 03

Required:
  0 = Optional
  M = Mandatory

a. NIF Announcements are redundant in SMT 7.2 because it is now mandatory to report Downstream Neighbor Address, which is not reported in NIF Announcements but in NIF Requests

8.7.4.2 Version ID. This is a one byte field which is used to indicate the version of the SMT frame-based protocol *not* the SMT standard. It is used to notify other stations on the ring (most notably management stations) of the version of the SMT *language* that a station speaks. This would allow the other station to recognize and interpret that version (if it can).

There are several de facto standards of SMT which have been implemented in the field. SMT 5.1 is the earliest such de facto standard. The other is SMT 6.2. Interestingly enough, both have the same version ID! This is because the changes were additions rather than modifications to existing protocols. Hence, the ANSI committee decided that a change in the version number was not necessary. It is only from SMT 7.2 that the version ID changed. SMT 7.2 is the first ANSI SMT standard; the others were draft proposed (dp) ANSI specifications.

After a great deal of debate, it was decided to maintain some level of *common speak* between all past, present and future SMT versions. This was achieved by making the version ID for NIF, SIF, and ECF a fixed constant ($0 \times 00\ 01$). The other protocols such as PMFs, SRF, and RAF will carry a version ID of two. The idea is to allow a minimum interoperability level amongst the different SMT versions so that adding new stations to an existing ring will not cause a disconnect in the SMT protocols. It should be noted that the different versions of SMT in no way affect actual ring operation.

The different versions of SMT affects the management stations whose complexity may be increased in order to manage a mixed (pre-SMT 7.2 and SMT 7.2) environment. The different versions do not affect the end stations.

8.7.4.3 Transaction ID. This is a 32-bit field which is used to tag each frame uniquely. This allows stations to match the request frames to the responses. Every SMT announcement or request frame generated has a unique 32-bit tag (transaction ID). A station building an SMT response always uses the transaction ID of the corresponding SMT request. The standard does not specify a mechanism for generating the transaction ID. Most implementations use some form of pseudo-random generation technique which ensures that the numbers are not repeated within a reasonable time period (say an hour). This algorithm can be reinitialized across discontinuities such as power-off.

8.7.4.4 Station ID. Every FDDI station has a unique 64-bit identifier. The lower six octets are a universally administered IEEE address in MSB order. The upper bytes are not specified in the standard.

What is the need for a station ID field?

In complex FDDI stations, having multiple MACs, a station cannot be identified by the MAC address (as there is more than one). Moreover, the station may have *hot-swappable* boards in which case a constant station ID is needed for unique identification of the station.

Why does it need to be 64 bits and not 48 bits?

Although the lower 48 bits constitute an IEEE assigned individual address, the upper two octets are implementer defined. These octets can be used for domain architecture of the FDDI network and local management purposes.

8.7.5   Why Multiple Protocols and Multiple Frame Types?

Why multiple SMT protocols? What do the multiple protocols do? Why not one SMT frame type? Why the need for multiple frame types? There are several reasons.

8.7.5.1   Need for Multiple SMT Protocols.   In FDDI network management, essentially two protocol-types are defined:

• Request-response protocols

• Announcement protocols

The Request-response protocols can be divided into two parts:

a. Network monitoring

b. Network management

Certain monitoring protocols are automatic and operate periodically (NIF-based Neighbor Notification protocol). A station desiring to monitor node and network behavior can do so by either passively monitoring network traffic or by sending Request frames and collecting the responses (SIF).

SMT also includes the ability to manipulate (configure) a remote station. These are special types of request frames called parameter management frames.

There are two kinds of announcement protocols; a periodic announcement which advertises a station's presence and identity on the network (e.g. NIF-based or SIF-based), and an event-driven announcement (SRF-based) triggered by some abnormal happening on the network.

8.7.5.2   Need for Multiple Frame Types.   We illustrated the various protocols which are defined in SMT. Each protocol has an associated frame type. But there are more frame types than protocols! The reason is backward compatibility.

Since the SMT standard has been eight years in the making, several versions of SMT have been in existence (most notably 5.1 and 6.2). Several companies who implemented to SMT 5.1 and 6.2 were interested in maintaining backward compatibility.

New concepts lead to new requirements which translates to new protocols. During the eight years while SMT was being defined, the FDDI MIB was
reworked extensively and expanded significantly. Network management
through traps became popular with SNMP and extensive remote MIB manipulation was seen as a necessary feature. In order to provide backward compatibility, the existing frames could not be modified which meant new frames
needed to be defined!

8.7.6 Neighbor Notification Protocol

Wouldn't it be nice to know your neighbors on the network? Wouldn't it
help to know if there is a duplicate station on the network? In the absence of
traffic on the network, wouldn't it be nice to verify that the MAC transmit and
receive paths are alive?

The Neighbor Notification protocol (NNP) supports all of the above functions via the use of the NIFs. Let us examine how each of the functions are
accomplished using NIFs.

In order to determine a stations' neighbors, a *new* station is required to
transmit a NIF Request (with FC = SMT_NSA) to a broadcast address as soon
as it becomes operational on the ring, informing the world of its identity. The
NIF NSA request also includes information about its upstream and downstream neighbors' address (UNA and DNA) which will be set to unknown initially. The first station downstream receives the frame, copies it, and sets the
A and C indicators. When it receives a token for transmission, it sends a NIF
SMT_INFO response to the station which sent the request. The response
frame includes information about this station as well as UNA (which is the
address of the sender of the request). The NIF response may include information about the DNA (if known, else set to unknown address).

In the meantime, the request frame traverses round the ring and the
rest of the stations may copy the frame (because DA = broadcast), but *do not*
respond to the frame. (See section 8.6.3 for discussion on SMT NSA and
SMT_INFO FC fields.)

This request-response protocol is executed by every station periodically.
This period is typically set to 30 seconds.

Let us examine how the NNP works for determining the addresses for a
three-station ring of stations A, B and C. Once one station issues a NIF NSA
then the identification process takes two full passes on the ring.

First pass

• Station A transmits NIF NSA Request: UNA = unknown;
  DNA = unknown.

- Station B receives NIF NSA Request. It sets UNA = A and transmits NIF SMT_INFO Response. UNA = A; DNA = unknown and it also transmits a NIF NSA Request: UNA = A; DNA = unknown.
- Station C receives NIF NSA Request. It sets UNA = B and transmits NIF SMT_INFO Response. UNA = B; DNA = unknown and it also transmits a NIF NSA Request: UNA = B; DNA = unknown.

Second pass

- Station A receives NIF SMT_INFO Response from B and sets DNA = B
- Station A receives NIF NSA Request from C. It sets UNA = C and transmits NIF SMT_INFO Response: UNA = C; DNA = B.
- Station B receives NIF SMT_INFO Response from C and it sets DNA = C;
- Station C receives NIF SMT_INFO Response from A and it sets DNA = A.

At this stage all stations know their neighbors.

Once the ring is operational, how would a station determine if there is a duplicate on the ring? While there are several methods, the NNP is a easy and standard mechanism to aid in duplicate detection once the ring is operational. If there is a duplicate address, a station could receive a NIF Response frame with DA = My source address, E indicator = Reset, A indicator = Set, and C indicator = don't care. The station would know there is a duplicate on the ring with the same address as its own if the destination address field has the same value as its own address. Once a duplicate is detected, the LLC services in the station that detected the duplicate are disabled until the duplicate address problem is resolved. During the time that the duplicate exists, SMT frame services remain enabled. It is interesting to note that if the two duplicates are adjacent to each other, the downstream station may not be able to detect the duplicate because it would strip the frame (source address match) and may not even copy the frame as described in section 7.6.1.3. The NNP consumes very little network bandwidth: less than 1% in the worst case.

The ability to build a map of the network is one of the features of a network management tool. The NIF frames can also be used to gather information about any station by sending a NIF SMT_INFO Request to the individual station address. The ring map can be built by putting the MAC into promiscuous mode where the MAC would copy all the SMT frames and examine all the NIF frames on the network.

8.7.7 Traps in FDDI

A management station needs to have an up to date image of the network in order to perform effective management. There are two basic ways of achieving this—poll each station on the network frequently or rely on each station to provide any information change. Both methods have their pros and cons. Polling frequently can lead to a significant portion of the network bandwidth being consumed by management traffic (not a very desirable scenario!). However, polling too infrequently can cause valuable information to be lost if multiple changes occur between the polling intervals. Relying on each station to provide information on any change can be an unreliable and unpredictable method. For example, if a node fails (reboots or hangs), it may not be able to inform the network manager about the problem.

8.7.7.1 Status Report Protocol.   To overcome the shortcomings of both methods, a combination of both protocols is implemented. In the combination protocol, any *unexpected* network or node change causes a status report frame (SRF) to be generated, informing the network of the unexpected change. The SRF contains information about the condition or event that triggered the SRF, a timestamp[3] of when the event occurred, and a transition timestamp of the time since the last unexpected event occurred.

What is the destination address used in the frame? Is it the address of the management station? Is it a broadcast address? Or is it a multicast address? Sending a frame to the management station requires every station to know the address of the management station (if any) on the network. Sending the frame to a broadcast address requires uninterested stations to copy the frame. To do this, SMT has allocated a special multicast address to which the frame is directed. Any station interested in observing the dynamic changes occurring in the network can monitor all frames with DA = SMT_SRF_DA.

What are the events or conditions?

An event is a *change* on the ring which is instantaneous and does not persist. A condition is an undesirable occurrence on the ring which is sustained. When the ring wraps due to a break in the cable, it is a condition because the *undesirable* topology persists until the problem is fixed. The wrapping of the ring causes a change in the neighbors for the nodes adjacent to the wrap point (figure 8.5); this is an event because the neighbor changed and is not necessarily undesirable.

Other examples of conditions are a large number of errors causing the Frame_Error_threshold to be equalled or exceeded or LER equalling or exceeding the LER_Alarm threshold, and if a duplicate station is detected then this is called a condition. Another example of an event is an illegal connection attempt (such as M to M ports).

---

3. The timestamp is a monotonically increasing value indicative of when the frame was generated since the last time the station was rebooted.

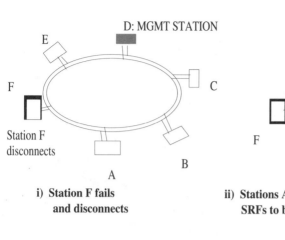

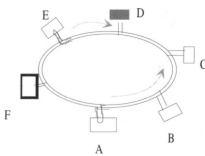

i) Station F fails
   and disconnects

ii) Stations A and E wrap. This causes
    SRFs to be generated by stations A & E.

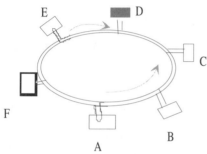

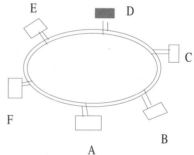

iii) Station D receives the multicast
     SRFs from stations A and E and
     pin-points the failed station.

iv) The problem is corrected
    and station F is back on
    the ring shortly at which time
    SRFs are again generated by
    Aand E indicating the change.

Figure 8.5 An example of status report frames.

8.7.8 Parameter Management Protocol

The ability to remotely monitor and manage a station is an important function of network management. Remote management is performed via frames. Typically other remote network management protocols (such as SNMP) have used the protocol stack PDUs (such as SNMP/UDP/IP). In SMT, the remote management is performed via a class of frames called the parameter management frames (PMF).

The basic functions of remote management include:

• Getting information from a remote station

• Changing attribute values in a remote station

• Deleting attribute fields in a remote station

In order to perform these functions, the following two PMF types are defined:

a. **PMF Get:** This frame class is used to obtain an attribute value, a group of attributes, or a range of attributes. A station issuing a PMF Get Request (of an attribute or a set of attributes) to a station receives a PMF Get Response from the station containing the requested values of the attributes (if attributes supported).

b. **PMF Set:** This frame class allows a management station to remotely configure a station. A management station desiring to configure a station first issues a PMF Get Request to obtain the current value of a given attribute. Upon acquiring the value of the attribute, it can issue a PMF Set causing the remote station to set the attribute value to the desired value. This is a very powerful command which, if used indiscriminately, can wreak havoc on the network. In order to provide a measure of security against indiscriminate usage of PMF Set commands (frames), several mechanisms are provided. In the first system, a Set_Count semaphore-type locking mechanism is used which gives a management station the ability to detect additional (possibly unauthorized) management stations. In the second system, an authorization field is used to exchange security codes (keys). These codes would be periodically changed in order to prevent unauthorized stations from causing any changes. The third mechanism, often the favorite choice, is not to implement PMF Set at all since it is optional!

8.7.9 Synchronous Bandwidth Allocation Protocol

The FDDI MAC essentially supports two classes of traffic: synchronous and asynchronous. The synchronous class of traffic is essentially guaranteed a pool of bandwidth at all times. The asynchronous class of traffic gets the remaining bandwidth. Most FDDI implementations today utilize the asynchronous channel only. The synchronous channel is very useful for stream-oriented traffic such as voice and video which requires a low latency, and high bandwidth pipe. As multimedia applications become a reality, the use of the synchronous channel becomes more important.

For the correct operation of the synchronous channel, it is necessary to allocate bandwidth to every station using the synchronous channel since every station can transmit up to the TTRT. This can cause the bandwidth to be overallocated and the TTRT protocol to fail resulting in frequent ring crashes.

8.8 FDDI MANAGEMENT WITH OTHER PROTOCOLS

FDDI is the first LAN to define its own frame based management (FBMT) for remote management. Several competing remote management protocols exist. The most popular amongst them is the Simple Network Management protocol (SNMP). SNMP was developed at the IETF and has become immensely popular amongst the management managers to manage everything from bridges, routers, and modems to address tables. SNMP is a frame-based protocol which can access remote MIBs and SNMP can manipulate individual attributes in a uniform manner. SNMP uses the frame services of UDP/IP protocols. This requires each managed node to have a UDP/IP protocol stack. While this is not a problem in a TCP/IP network (such as workstations using BSD4.3, System V UNIX, SunOS 4.1, etc.), it can be inconvenient in a mixed environment where some stations are using the TCP/IP protocol stack, some are using Novell Netware, and still others using AppleTalk.

The IETF has developed a SNMP-specific MIB for FDDI. The SNMP MIB is based on the FDDI SMT MIB and is very similar. The reason for developing a separate SNMP MIB is that SNMP uses a different subset of the ASN.1 datatypes. Several attributes and attribute groups defined in FDDI use datatypes which are not defined in the SNMP definition. The IETF has translated such attributes into SNMP compatible format. Most FDDI vendors who supply an SNMP MIB for their FDDI system (adapter or concentrator) are compliant to the relevant FDDI SNMP MIB RFC.

ISO has defined a remote management protocol called the Common Management Information Protocol (CMIP) to which a lot of companies have paid lip service but never actually implemented. This protocol, although very powerful is extremely cumbersome and has never been fully implemented. This protocol has not seen much use.

There was significant resistance amongst the FDDI community to implementing a similar FDDI-specific protocol such as PMF when SNMP was widely available and used to manage different networks. The argument against PMFs has been that it is unnecessary to define an FDDI-specific protocol when existing protocols such as SNMP and CMIP could be easily tailored to meet FDDI requirements. The arguments for the PMFs have been its protocol independence, small size and extensive MIB manipulation abilities.

Today, most FDDI vendors provide SMT FBMT and SNMP management. Table 8.3 shows a comparison of SMT FBMT, SNMP, and CMIP. A station implementing FBMT, SNMP, or CMIP can share the same MIB as shown in figure 8.6.

Table 8.3   Comparison of SMT, SNMP and CMIP

Description	SMT	SNMP	CMIP
Standards body	ANSI X3T9	IETF	ISO
Protocol stack requirement	None	UDP/IP; other protocols available	OSI ROSE, FTAM, TP-4, etc.; others are CMOT
MIB format	SMI	SMI	SMI
Language of MIB description	small subset of ASN.1	small subset of ASN.1	larger subset of ASN.1
Encoding rules for attributes	Combination of BER and others	BER	BER
Mechanisms for MIB access:			
a) Fetch Attribute	GET; SET, CHANGE	GET; GET-REPLACE	GET; SET, GET-REPLACE
b) Change			
c) Action	ACTION	--NA--	ACTION
d) Create	--NA--	--NA--	CREATE
e) Delete	--NA--	--NA--	DELETE
f) Initialize/Terminate	--NA--	--NA--	INITIALIZE/ TERMINATE
g) Multiple Attribute fetch	MULTIPLE GET	GET NEXT	GET with SCOPE/FILTER
Trap feature	Yes. SRF	Yes. Trap PDUs	Yes. EVENT-REPORT
Decoding of frame content	By table: first octet is Type: A given type has entry in table specifying data type. No implicit info. of data type in frame	By algorithm: first octet contains info. on data types followed by object type, length and data. Table not required	By algorithm
Size of frame containing n attributes	small	large	larger
Local subnet management	SMT Frame based Management suffices	SNMP or by SNMP proxy	OSI stack or proxy
Enterprise management	--NA--	SNMP and proxies	OSI and proxies

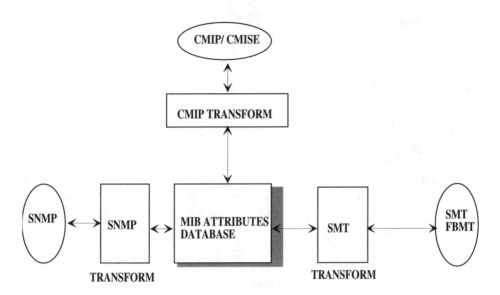

Figure 8.6   MIB translation across different protocols.

Internetworking with FDDI

9.1 INTRODUCTION

A typical enterprise network spans large geographical boundaries and consists of LANs, MANs, and WANs. A LAN is usually restricted to a few kilometers in span. LANs were initially targeted for PC to mainframe and PC to peripheral connectivity. Workstation to workstation connectivity has become increasingly prevalent which has led to a proliferation of networks in the corporate office environment and the development of *enterprise* networks. Enterprise networks consist of many different LANs within campuses and wide area connections (such as T1/T3 lines) between campuses. Such a heterogeneous enterprise network spanning LANs and WANs is called an *internetwork*.

In an internetwork, sometimes the subnetworks are different (e.g., serial lines, Ethernet, FDDI), and sometimes the protocol stacks are different (AppleTalk, Novell IPX, TCP/IP, DECNET, SNA, etc.). The LANs, WANs are interconnected with bridges, routers and gateways, also known as *interworking units (IU)*, (figure 9.1.) For reasons of security, administration and ease of

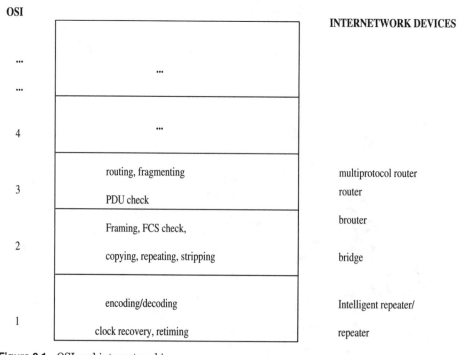

Figure 9.1   OSI and internetworking.

management, networks are *segmented*. These segments are connected together via devices which offer traffic filtering and isolation. Therefore, even in a homogenous network (similar protocol stacks and/or similar subnetworks) interworking units are used.

9.2 REPEATERS

A repeater is a physical layer IU which amplifies and retimes the incoming signal and repeats it. Repeaters are useful in applications where it is important to extend the maximum distance between nodes within a LAN or WAN at a low cost. The disadvantage of using a repeater is that it will propagate most errors thus compromising link management abilities. In FDDI, it is possible to have repeaters with additional capabilities. A repeater consisting of PHY devices not only amplifies and retimes the incoming signal, it also performs line-state signalling and error detection. Typically, these are available in a multiport configuration and are called MACless concentrators. MACless concentrators (repeaters) provide better link management and error detection capabilities than a basic repeater at an incrementally higher cost. MACless concentrators are theoretically disallowed in the SMT standard (see configuration management section in SMT) although many low-cost solutions implement them anyway.

A different type of repeater is useful in environments having 50 and 62.5 micron cable which need to be connected together in an FDDI LAN environment. Several FDDI vendors provide *mixed port types* where the A port may have 50 micron transceivers and the B port may have 62.5 micron transceivers. Another application where these repeaters are useful, is in a network where nodes are very far apart (as in different buildings). For this kind of applications, the user should carefully evaluate the cost-performance tradeoffs of transceiver repeaters versus concentrators. Transceiver repeaters may be used if low-cost is the main criteria. Otherwise, it is recommended to use a mixed concentrator with ports supporting different media (e.g. multimode fiber and single mode fiber) rather than a mixed set of transceivers.

Repeaters are not well described in the FDDI standards. This lack of standardization makes it difficult for users to provide a standard repeater specification to their vendors. As a result, different vendors have different methods of implementing repeaters which can sometimes lead to interoperability problems.

9.3 BRIDGES

Bridges are used to extend network segments or to interface different subnets (such as Ethernet and FDDI). Bridges are IUs which deal with layer 2 (data link layer) addressed frames and connect two or more LANs. They are also known as *MAC-level relays*, and *link extension gateways*. A bridge has a minimum of two *ports* where each port is a connection to a different LAN. In a bridged environment, it is necessary that each individual station on a bridged network has unique addresses. Bridging, at a minimum, implies performing address filtering and forwarding of appropriate frames. A bridge has several desirable properties such as traffic isolation, increased network span, increased number of stations, multiple media (e.g. baseband coax, twisted-pair, fiber), interconnection of dissimilar LANs, high throughput, low delays, and topology flexibility.

9.3.1 Types of Bridges

Bridging can be broadly categorized as shown in figure 9.2. Bridges can be remote or local. Remote bridges typically connect a LAN to a WAN and rely on a symmetrical arrangement at the other end of the connection, (figure 9.3.) The remote bridges may run proprietary protocols or better still, protocols such as PPP (Point to Point protocol) or SLIP (Serial Line Internet protocol) in order to extend the LANs seamlessly across the WAN. Local bridges are faster, more efficient, and have less protocol overhead than remote bridges. Typical remote bridge links are analog leased-lines (up to 19.2 Kbps), digital lines (56 or 64 Kbps) or fractional or multiple 1.544 T1 links. Although FDDI can be bridged remotely, most FDDI remote bridges will drop a substantial number of packets if a burst of packets arrive at the bridge at 100 Mbps to be forwarded at 19.2 Kbps! When buying a remote FDDI bridge to a slow link, it is important to ensure that the bridge has large buffers to prevent packet drop-

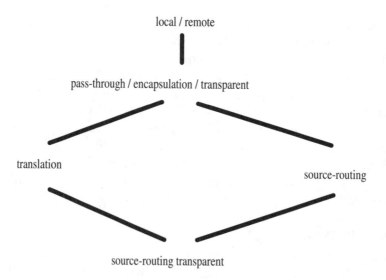

Figure 9.2 Types of bridges.

ping. Packet sizes may need to be smaller to prevent higher-layer protocol time-outs. It is envisioned that FDDI will be bridged remotely by SONET links at 55 Mbps or 155 Mbps.

9.3.2 Local Bridging

There are three kinds of local bridges – pass through, encapsulation, and translation.

These bridges forward frames based on one of two mechanisms– transparent bridging (which is the IEEE 802.1d recommended method of bridging) and source routing bridging (which is the IEEE 802.5 recommended method of bridging, although source routing can be applied equally to other LANs). After years of recommending either of the two mechanisms, the IEEE 802.1d committee adopted transparent bridging for IEEE LANs. In order to accommodate the existing source routing bridges, a compromise *source routing transparent* (SRT) bridging was adopted by the IEEE 802.1d committee. In SRT, a transparent bridge will not forward a source-routed frame and a source-routing bridge will not forward a non-source-routed frame.

9.3.2.1 Source Routing Bridging. In source routing, all end-stations maintain a routing table and a list of the location of each end-station. In source routing the end-stations are more complex and the source routing bridges are simpler than the end-stations and bridges in transparent bridging. In SRT both kinds of stations and therefore both kinds of bridges can coexist on the same LAN. The setting of the routing information indicator (RII) bit in the information field of the frame indicates if the frame is to be source routed (bit set), or transparently bridged (bit reset).

9.3.2.2 Transparent Bridging. In transparent bridging, the complexity is in the bridge and hidden from the end-stations. The end-stations are unaware of the location of the destination station which may be on the same LAN or it may be across the bridge. The bridge maintains a *spanning tree* of all the bridges in the extended LAN and forwards the frames transparent to the sending end-stations. A spanning tree is a path list of one and only one path to all the nodes in an extended LAN (multiple LANs connected by bridges). The path list avoids logical loops by ensuring one and only one route to any node. The spanning tree algorithm is necessary because logical loops can lead to packets circulating endlessly around the network. The spanning

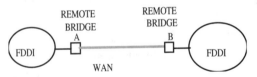

Figure 9.3   Remote bridging.

tree protocol for bridges was first defined by DEC and later standardized by the IEEE 802.1d committee with minor modifications.

In transparent bridging, a bridge *learns*, *filters* and *forwards*. There are a number of different algorithms for learning, filtering and forwarding. One such algorithm follows:

- At initialization, a bridge is *learning* about other stations on the network. During initialization, every packet on the wire at each port is examined by the bridge and its destination address is copied into each port's forwarding table.

- Each packet is examined and the source address is checked in the forwarding table for that port. If there is no match in the table for the port, the address is added to the table. If there is a match, the time-stamp is reset indicating that the entry is not outdated. If there is a match on a different port (indicating that the host moved from one segment to another), the entry is changed to indicate the new location and the time-stamp is reset.

- The destination address is then checked against the forwarding table of other ports and forwarded appropriately or if there is no match, it is broadcast to all ports and added to the destination address field in every table. During the first few minutes after initialization, the bridge is in learning state only and does not forward any frames.

- Upon completion of initialization, the frame is ready to filter, forward, and learn. Filtering is the process of examining a packet in order to determine whether to forward or discard the packet. Forwarding is the process of delivering a packet to the appropriate port (changing headers, regenerating checksum, copying to the appropriate transmission queue). A multi-port bridge will forward the frame to the appropriate port based on a destination address match. If it finds a match on a port, it forwards the frame to that port. If it matches the destination address on the incoming port, it discards the frame. If no match is made, the frame is forwarded to all ports (except the incoming port). This process is called *flooding*.

Some vendors provide the ability to apply custom masks on the frames in order to restrict traffic to local LANs. This additional filtering can prove to be useful in a completely bridged environment where there is otherwise little scope for traffic control. A time-stamp on all entries provides a bridge with the ability to remove entries which have not been active for a given period of time. This process is called *aging of table entries*. This allows the forwarding table to be pared in a least recently used (LRU) fashion.

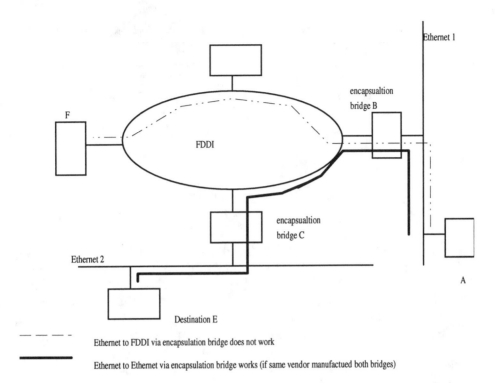

Figure 9.4   Encapsulation bridging.

9.3.2.3   Pass-Through Bridging.   A pass-through bridge is only possible if the bridged LANs have identical frame formats (such as FDDI and Token-Ring) or are identical (such as two Ethernet networks). A list of the network (or segment) location of each station is kept in the forwarding table (typically a content addressable memory). The destination address field of the incoming frame is checked against the forwarding table and a determination is made whether to forward the frame to another LAN or to discard it. If there is an address match on the local ring, the frame is discarded. If there is an address match in the address database of a port, the frame is forwarded to that port. If the frame is a broadcast frame (or no destination address match could be made on any port), it is forwarded (or flooded) to all ports of the bridge (networks) except the incoming port. In pass-through bridging, a frame may be forwarded *before* it is completely received. This is also known as *cut-through* bridging and it increases the throughput rate of the bridge. These bridges are very popular in Ethernet-Ethernet bridging.

Pass-through bridges cannot be used in FDDI-Ethernet bridging because of the differences in frame formats and also the transmission order of the bits and bytes. These bridges are usually the fastest and simplest because of the simplicity of the transform algorithm.

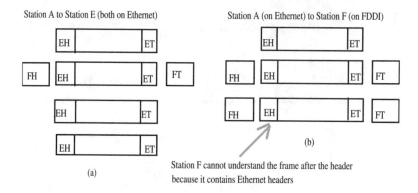

Figure 9.5   Encapsulation bridging: Ethernet to Ethernet (a) and Ethernet to FDDI (b).

9.3.2.4   Encapsulation Bridging.

An encapsulation bridge wraps the frame from the incoming LAN in the outgoing LAN's headers and trailers and forwards it on the intermediate LAN. The receiving bridge on the intermediate LAN unwraps the headers and trailers and forwards the original frame to the destination LAN. This method was favored by several FDDI vendors at one time. However, this technique has now fallen out of favor because of inherent limitations–this works well if the encapsulation bridge is used in a backbone (e.g. FDDI) where the end-stations are on a bridged LAN (Ethernet) and the end-to-end communication is from a station on the same LAN. It falls apart when the sending and the destination stations are on different networks. For example, if the frame from station A (on Ethernet) is encapsulated by the encapsulation bridge and transmitted on the FDDI network with the address of station F (on FDDI), station F will not be able to decode the contents of the encapsulated frame (the Ethernet headers) because it is not aware of the encapsulation.

9.3.2.5   Translation Bridging Issues.

A translation bridge is required if the subnetworks are different (Ethernet and FDDI) but protocol stacks are identical (TCP/IP). The translation bridge translates the format of one LAN to another. Issues of concern to a translation bridge are network speed mismatches (10 Mbps v/s 100 Mbps), network PDU size mismatches (1,500 bytes vs. 4,500 bytes), network address format mismatches (little Endian vs. big Endian), frame priority handling (e.g. synchronous and asynchronous frames), CRC regeneration, and source-routing information. Some of the issues in translation bridging from Ethernet to FDDI and vice versa are shown in figure 9.6.

9.3.2.5.1.   IEEE 802.3, Ethernet and FDDI.

It should be noted that the frame formats for the Digital Intel Xerox (DIX) Ethernet are different from the IEEE 802.3 CSMA/CD frame formats. Because of this difference, an Ethernet/FDDI bridge is not the same as an IEEE 802.3/FDDI bridge. 802.3 uses the 802.2 frame formats which are similar to FDDI frame formats: SD-{FC}-DA-SA-DSAP-LSAP-CONTROL-INFORMATION FIELD-FCS-ED. When using IP over any 802 LAN, (see RFC 1042), or IP over an FDDI LAN (see RFC 1188), the LLC address fields, destination service access point (DSAP) and source service access point (SSAP), are set to a Subnetwork Access protocol (SNAP) ID. This constant has been defined in IEEE 802.2 as a byte value equal to 170 decimal (10101010 binary or AA hex) to be used for encapsulating different protocols. The presence of the SNAP value indicates that the five bytes following the LLC control information field are to be treated specially. The first three octets are a special organizationally unique identifier (OUI) value ($0 \times 00\ 00\ 00$) followed by 16 bits of the EtherType (or protocol ID) as listed in the RFC 1010 (assigned numbers). A bridge or router uses these fields for de-multiplexing and forwarding the packet correctly.

9.3.2.5.2.   Subnetwork   Access   Protocol   and   Translation Bridging.   The SNAP scheme was formulated because the IEEE 802.2 headers did not have a standard mechanism for de-multiplexing different protocols. The DIX Ethernet specification defines a *type* field to indicate the protocol type. While the SNAP encapsulation works well with 802 LANs, when bridging between DIX Ethernet and IEEE 802 LANs, a bridge has to perform the SNAP to Type field translation and vice versa. This translation is unfortunately complicated when one of the LANs involved has the AppleTalk protocol stack. Apple-

Ethernet to FDDI

- Address field reordered
- Type field removed
- replace Ethernet preamble with FDDI
- SNAP/UI/P_ID added
- padding is removed
- recalculated FCS appended

FDDI to Ethernet

- Address fields reordered
- P_ID= Special OUI + Type field
- Replace FDDI preamble with Ethernet
- padding inserted if needed
- recalculated FCS appended

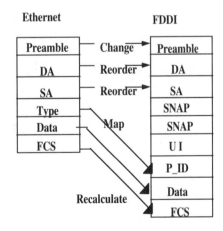

Figure 9.6   Ethernet to FDDI bridging.

Talk supports DIX Ethernet (Phase I) and IEEE 802.3 (Phase II) networks. It uses the OUI of $0 \times 00\ 00\ 00$ in its Phase II packets. A translation bridge seeing a Phase II AppleTalk packet would convert it to a Phase I packet format because the bridge believes that it is a SNAP encapsulated packet, even though the destination is not a Phase I station or network. This can be resolved in two ways–Route AppleTalk and not bridge it, or DEC offered an alternative solution to flag a Phase II packet. It offered an OUI of $0 \times 00\ 00\ F8$ (with permission from IEEE) as a flag for Ethernet packets to be translated back to Ethernet packets by intervening non-Ethernet bridges. The OUI of $0 \times 00\ 00\ 00$ would be used with Appletalk Phase II. Thus, $0 \times 00\ 00\ F8$ is the Ethernet Encapsulating OUI which is used by bridges to relay Ethernet frames across non-Ethernet or IEEE 802.3 LANs.

9.3.2.5.3. Ethernet to FDDI Bridging.

Bridging from Ethernet to FDDI is easier than going from FDDI to Ethernet for two reasons:

- Mapping smaller frame sizes to larger frames implies no segmentation.

- Going from a slower to a faster network implies minimum buffering.

Of course, the same two reasons make it difficult to bridge from FDDI to Ethernet. Because of the speed mismatch, if a burst of frames arrive at the bridge, the bridge has to either provide a very large buffer or drop some of the frames. An FDDI-Ethernet translation bridge not only does the functions mentioned above, it also has to grapple with other more sticky issues.

What if the FDDI frame to be bridged is $> 1,518$ bytes?

What if the packet size is < 64 bytes (as may be the case with an ICMP *ping*)?

An FDDI to Token-Ring translation bridge is slightly easier to build because of similarities in frame formats (and sizes) and the bit order of the address fields but the network speed difference is still a problem.

9.3.2.6 FDDI Bridge Frame Stripping.

An important issue in translation bridging in FDDI is frame stripping. Frame stripping is done because each station is responsible for removing frames that it transmitted. If the frames are not stripped, the same frames will be received by the destination stations repeatedly. This causes the receive buffers to overflow and can lead to packet dropping. A station strips a frame by checking the source address on every frame and stripping it if there is a match[1]. A bridge cannot strip forwarded frames in a similar fashion because the source address on the forwarded frames is not the bridge's but of end-stations which are not on this

FDDI ring. Over the years, several mechanisms have been proposed and implemented:

- Maintain the source address of each forwarded frame and use *external source address matching*. Internal source address matching is when the MAC can match the source address on the incoming frame with the source address of the bridge (or station). External source address matching is when the MAC is provided an external indication that there is a source address match and hence should strip the frame. The source address of the frame is matched with a list (or table) of source addresses of the forwarded frames. Although this is very effective, it is not a desirable solution. It requires additional hardware and can be expensive in terms of the size of CAM required.

- Count the number of frames forwarded and strip an identical number of frames (counter method). Although this method is fairly simple, it can lead to *over-stripping* (i.e. stripping more frames than the actual number forwarded). If the counter increment/decrement method is accurate, this method is not susceptible to *under-stripping* (i.e. stripping less frames than the actual number forwarded). If a forwarded frame is lost due to corruption, the bridge will strip one more frame then it should have. If the FDDI network is hit by a burst of noise corrupting several frames, it is possible to strip frames sourced by other stations on the LAN. This can disrupt network services.

- Mark the end of a block of forwarded frames and strip all frames until the marker is reached (marker method). A MAC typically implements this by sending one, two (or more) fully-formed void frames at the end of the forwarding *before* releasing the token. This scheme is susceptible to over-stripping if the marker is corrupted and cannot be recognized as such. It is also susceptible to under-stripping if a false marker is created due to frame or inter-frame gap corruption.

- A combination of the counter and marker method. This method has a low probability of under-stripping as provided by the counter method and a low probability of over-stripping as provided by the marker method. The counter increments for every frame forwarded and decrements for every frame received. At the end of the forwarding a void frame (DA = SA = My SA), is transmitted. Every frame received is stripped until the counter reaches zero or the marker (My void frame) is received. This scheme was proposed by DEC and is known as *void*

---

1. Some frames are automatically stripped when a station is transmitting. This occurs when a station transmits sufficiently longer than the ring latency.

stripping algorithm. Motorola and AMD chipsets implement this or a modified form of this algorithm.

9.3.3 Issues in Selecting FDDI Bridges

In selecting an FDDI bridge, the user must consider:

9.3.3.1 Bridge Type. As discussed earlier, there are several types of bridges. In selecting a bridge, a user is cautioned against purchasing FDDI devices that are vendor-specific. An example of a proprietary vendor-specific bridge is an encapsulation bridge. A user should buy an industry standard transparent bridge which is compatible to the IEEE 802.1d, IEEE 802.1i (FDDI supplement to the IEEE 802.1d bridging standard), and IEEE 802.1h. The IEEE 802.1i defines bridging between FDDI and any IEEE 802 LAN. The IEEE 802.1h was established to support non-standard protocols such as AppleTalk Version 2.0 which is non-compliant to the IEEE 802.1d/802.1i standards. If a user has installed token-ring networks and is using source-routing bridging, the user is well-advised to acquire source routing transparent (SRT) FDDI bridges.

9.3.3.2 Network Management Support. Some bridges include a rich set of management features including SNMP and SMT PMFs. SNMP is a preferred method of network management although SMT PMFs are a powerful low-level FDDI-only management tool. If installing a new FDDI network, ensure that the latest version of SMT is used. Full FDDI remote FBMT (including PMF Get Requests) is not necessary and if offered by the vendor, the premium should be negligible (less than 10%) and SNMP management should be the first choice.

Bridges with security features to prevent unauthorized users from modifying bridge databases or the extended LAN topology are strongly recommended.

9.3.3.3 Frame Filtering. Frame filtering is the rate at which the bridge can pick frames off the wire. This is not as useful a measure as forwarding rates. However, frame filtering rates less than 2,778 frames per second are unacceptable. This is the network rate for 4,500 byte frames at 100 Mbps. High numbers imply that the bridge has a very fast backplane and table look-up algorithm.

9.3.3.4 Forwarding Rates. Forwarding rate is the rate at which the bridge can take a frame from one network and transmit it onto the appropriate network. Each vendor uses different packet-sizes and different performance parameters (peak, sustained or average throughput) to specify their bridge and therefore when buying a bridge it is important to compare similar

packet sizes and performance parameters. The peak and sustained throughputs provide the best information about a bridge's behavior under bursty traffic (short periods of heavy data movement) and under sustained load
(prolonged periods of heavy data movement).

9.3.3.5 Frame Sizes. An FDDI-Ethernet bridge may support frame
sizes of 1,500 bytes only. This may be acceptable if the bridges form a bridged-
backbone environment with all stations on the backbone being bridges or routers. If however, the user intends to place an end-station (a workstation, server
or mainframe) on the FDDI LAN, it is desirable to have bridges that support
segmentation and reassembly (SAR). Such bridges typically incorporate some
network layer functionality (e.g. SAR), but may not be full-fledged routers
(performing ES-ES routing). Other choices may include routers and brouters
(route if you can, bridge if you cannot). A bridge which does not support frag-
mentation can dramatically impact the performance protocols such as NFS. In
such an environment where machines are capable of high throughput, it is
preferable to have a brouter.

9.3.3.6 Frame Priority. It is desirable to have bridges with the ability
to forward higher priority frames (such as synchronous or asynchronous priorities). Bridges offering a minimum delay path to synchronous traffic (e.g. multimedia streams) are desirable. Currently, most bridges do not distinguish
frames based on frame control fields (other than to filter out MAC, SMT and
void frames).

9.3.3.7 Load Balancing. FDDI bridges should maintain a spanning
tree and not introduce loops in the bridged environment. In a spanning tree
environment, only one path exists to a destination. Some FDDI bridges maintain more than one path to a destination and are capable of exploiting loops in
the topology to distribute the load across the two paths. The ability to share
load on multiple low-capacity links (e.g. remote FDDI bridge links) is a defi-
nite plus.

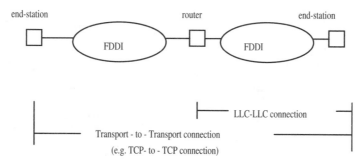

Figure 9.7 Routing.

9.4 ROUTERS

Routers operate at the network layer (OSI level 3) and perform the routing function on behalf of other network nodes. Routers are used when interfacing to different LANs either locally or remotely (across a WAN) and provide flow control and traffic isolation to the different network segments. While a bridge is transparent to the LLC connection, a router is an end-point of the LLC connection, (figure 9.7.) In other words, an end-station is aware of the address of a router and specifically directs frames to the router to forward to the destination station. In bridging, the end-station is not aware of the presence of a bridge. The router maintains a database of the network addresses and the status of the network. It maintains the addresses of other routers and directly reachable end-stations. Both types of information are maintained by periodically exchanging information with other routers and reachable end-stations. There are several protocols used to exchange the information and maintain a *shortest path* table. These protocols are called routing protocols and some of the more popular one's are Open Shortest Path First (OSPF), Intermediate System-Intermediate System (IS-IS), and Routing Information protocol (RIP). Most of these routing protocols are based on the algorithms developed by Bellman-Ford, Djikstra or Floyd-Warshall.

9.4.1 Single Protocol Routers

Like routers use the same network layer and hence the same routing protocol. Such routers are single protocol routers which may offer subnetwork (MAC addresses) but not protocol independence because they understand one protocol only and use only one routing protocol. An example of single protocol routers is an IP router with Open Shortest Path First (OSPF) routing protocol.

9.4.2 Multi-Protocol Routers

Multi-protocol routers can route many different protocols such as DECnet, TCP/IP, SNA, XNS, Netware, and other environments, (figure 9.8.) A

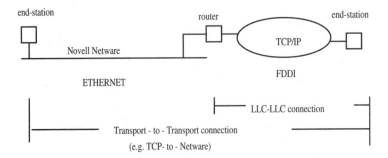

Figure 9.8 A multi-protocol router.

multi-protocol router maintains separate databases for each protocol that it supports. There are many ways of performing multi-protocol routing of which the three common methods are Ships In the Night (SIN), Integrated Routing, and Encapsulated Routing.

9.4.2.1 Ships In the Night (SIN). As the name implies, each protocol is routed independently of the other protocol. For each protocol to be routed, there is one routing protocol. As an example, a SIN multi-protocol router implementing DECnet, TCP/IP, and OSI would use DECnet to route DECnet, OSPF (in newer implementations) or RIP (in older implementations) to route TCP/IP and IS-IS to route OSI. SIN routers maintain separate virtual networks for each protocol over the same extended physical network. This task can be fairly complex to implement and manage. Due to the additional protocol processing, a SIN router generally has lower throughput than single protocol routers while affording maximum connectivity between disparate networks and protocols.

9.4.2.2 Integrated Routers. Due to the complexity of SIN routers, many router vendors have developed integrated routers which use the same routing protocol to route multiple protocols. For this purpose, they have developed an Integrated IS-IS protocol which performs the task of routing across different protocols. This method of routing has the advantage of reducing the multiple databases to one and simplifying the management of the physical and virtual network, while at the same time affording multi-vendor interoperability.

9.4.2.3 Encapsulating Routers. Encapsulating routers are routers which encapsulate the different end-protocols into one common intermediate protocol for the purposes of routing. For example, DECnet, SNA and XNS would be encapsulated into IP for routing purposes and a single routing protocol such as OSPF would be used. While this scheme has the advantage of a single database and a single routing protocol, it has the same disadvantages that an encapsulating bridge has. One major disadvantage is that the proprietary encapsulation schemes lock the user to the same vendor for any internetworking devices.

9.5 BRIDGE AND ROUTER: BROUTER

Numerous vendors now offer hybrid products which combine the best features of bridging and routing: speed and traffic filtering. These products are called *brouters,* (figure 9.9.) The concept is to "route if you can and bridge if you must". Several people have argued otherwise and believe "bridge if you can and route if you must". While both arguments have their merits, the resultant products can differ substantially.

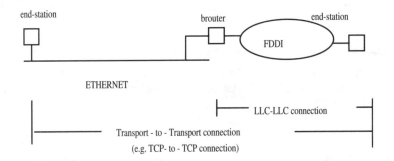

Figure 9.9   Routing and brouting.

In the *route-if-you-can* paradigm, the per packet processing increases substantially as compared to a bridge. If a packet arrives at the brouter, and after processing the headers, it is determined that the packet-size and protocol is supported by the destination LAN, the packet may be bridged. This avoids further processing and a higher throughput is achieved. This processing may be speeded up by custom hardware or very fast CPUs.

In the *bridge-if-you-can* paradigm, the underlying assumption is that the upper-layer protocols are identical and only subnets are different. Thus, all packets are primarily bridged, except for packets larger than MTU of destination LANs, no match for forwarding, and so on. Under these circumstances, the packet is passed to the network layer for further processing.

9.5.1   The Trade-Offs

- Bridges are cheaper than routers
- Bridges provide higher throughput in homogeneous environments (same physical subnetwork) because of minimal processing and store-and-forward overhead.
- Routers offer better throughput than most bridges in heterogeneous environments with different subnetworks such as Ethernet, FDDI and Token-Ring. Routers allow the use of maximum sized MTUs on each interface which increases the network efficiency and throughput. Thus end-stations on FDDI can used maximum-sized frames (4,492 bytes) to improve the FDDI throughput. If the destination is on an Ethernet across a router, the router segments the FDDI MTU into smaller Ethernet frames before forwarding. With a router, all traffic destined outside the LAN is *segmented and reassembled* by the router and local traffic can operate at optimum packet size.

- In a bridged environment, all stations transmit using the smaller of the MTUs of each LAN. Otherwise the bridge would have to drop packets larger than smallest MTU. In an FDDI-Ethernet bridged environment, all stations, including FDDI stations, transmit using 1,500-byte MTUs. This reduces the efficiency and throughput on the FDDI ring. An even more important impact is on the throughput of upper-layer protocols such as NFS over UDP and TCP transfers.

- If security and flow-control are important, a router offers better flow control, traffic isolation, monitoring and control facilities than a bridge.

- A multiprotocol router is necessary in a multiprotocol environment. A bridge cannot operate above the MAC layer and is unsuitable in mixed protocol environments where a translation between the different protocols is to be performed.

Cabling and Layout

10.1 INTRODUCTION

One of the least glamorous issues in computer networking is the underlying connection mechanism: cables (fiber, copper, or...). In any discussion on computer networking, it is very easy to assume that the wiring will be available and it is just a matter of connecting the computers. This model worked when computers were few and connecting computers to printers or mainframes was accomplished by *ad hoc* schemes such as tossing cables across cubicles! Ad hoc wiring is typically temporary, over the wall, and cable as needed. The ubiquity of the desktop computer and the variety of networks has forced users and vendors alike to resort to *structured* wiring. Structured wiring is typically more permanent, behind the wall, available at every work area, and has wall outlets (see figure 10.1).

Today's user faces a myriad of issues:

- Various networks (telecommunications, LANs, and backbones) (figure 10.2)
- Different applications (voice, data, and video)
- Multiple environments (office floor, building wiring, intra-campus wiring, and factory floor)

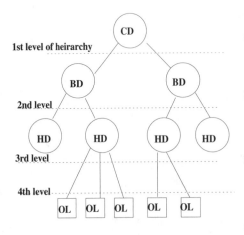

CD :: Campus Distributor. A collection of functional elements (e.g. patch panels, regenerators,) used to connect between campus points

BD :: Building Distributor. A collection of functional elements (e.g. patch panels, multiplexers,) used to connect between the building and campus cables

HD :: Horizontal Distributor. The connection point between the cabling to the telecom outlet and the cabling between the distributors

OL :: Outlet. The actual wall outlet where the heirarchical cabling terminates

Figure 10.1   Hierarchy of distribution of cable plants.

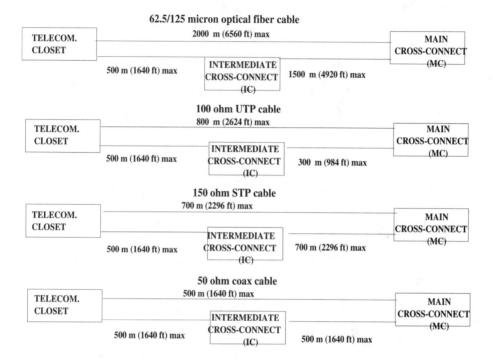

Figure 10.2 Cable distances in the backbone.

These issues cannot be handled in an ad hoc manner because today's office campuses are far more complex. For instance, IEEE 802.3, IEEE 802.4, and IEEE 802.5 specify different topologies, different distances, and different media and connectors. Providing all three within a campus can be a cabling nightmare. Structured wiring is standards driven and offers a comprehensive solution to cabling for a multi-vendor and multi-product environment, (figure 10.3).

10.2 CABLING FOR FDDI

The following issues should be considered when cabling for FDDI:
• The FDDI application (backbone or desktop)
• Existing networks and cable plants
• FDDI and wiring topology
• Flexibility and scalability
• Cost

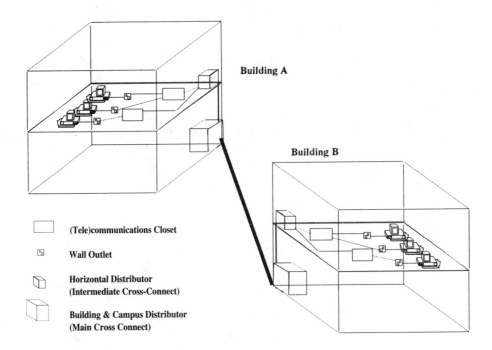

Figure 10.3 An example campus-wide cabling structure.

10.2.1   Designing an FDDI Backbone

FDDI is used in the backbone network application for three primary reasons. It provides:

• Higher bandwidth

• Longer distances between nodes and larger network span

• A secure, fault-tolerant network

A backbone network typically spans several floors and often across buildings and is used to tie existing networks together. A backbone network is strategically important to a corporation because a failure in the backbone can lead to significant communications breakdown and loss of productivity. Therefore, it is necessary that a FDDI backbone network maintain a very high up-time. Providing high up-time may require redundant cables and connections and a secure area for the FDDI backbone interconnection devices. There are at least two distinct ways of implementing an FDDI backbone with fault-tolerance and redundancy which are in accordance with the EIA/TIA 568 wiring standard:

• The backbone is formed by the DAS connections of bridges and routers. This topology is shown in figure 10.5a. In this topology, the DAS con-

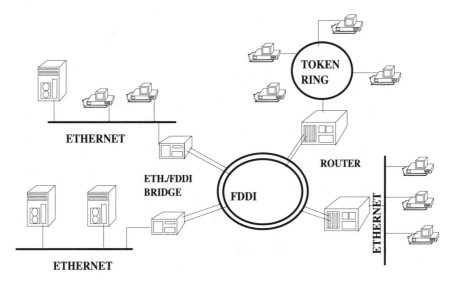

Figure 10.4a An FDDI backbone supporting multiple Ethernet and Token-Ring networks via bridges and routers.

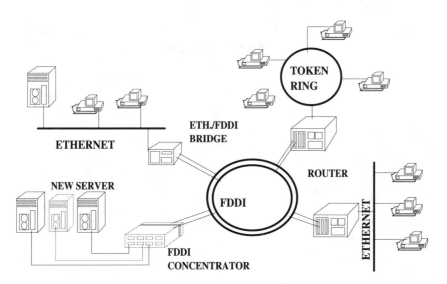

Figure 10.4b With the additional server, upgrade the server subnet to FDDI; replace the bridge with a concentrator. The rest of the backbone remains untouched.

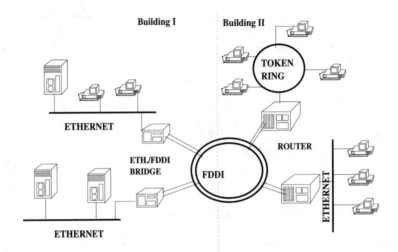

Figure 10.5a An FDDI backbone supporting multiple Ethernet and Token-Ring networks via bridges and routers across different buildings.

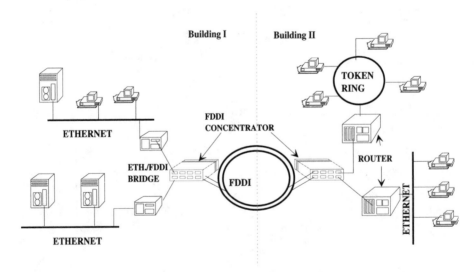

Figure 10.5b An FDDI backbone distributed through concentrators and bridges and routers. This topololy provides a scalable and fault-tolerant architecture.

nections should have an optical bypass switch in order to prevent the dual-ring from wrapping to a single ring if a station fails. There are several problems with this topology:

a. DAS without bypass switches can cause ring segmentation and isolation if more than two stations are disconnected.

b. DAS with bypass switches affects the FDDI power loss budget and maximum distance between nodes. Moreover, if more than four consecutive bypass switches become active, the ring may become inoperational or unreliable.

c. A DAS implementation implies that the user is directly connected to the backbone and thus reduces the fault-tolerance and fault-isolation capabilities of the backbone. For example, a user unplugging his DAS connection can cause the ring to wrap. This condition can only be overcome by re-configuring the cable plant.

• The backbone is connected in a dual attached ring of concentrators with a hierarchy of devices such as bridges and routers available off the concentrators as SAS connections. This topology is shown in figures 10.4b and 10.5b and is the recommended method of configuring FDDI networks as it avoids the problems with direct dual-attachment bridges and routers and provides the ability to add stations to the FDDI later without disrupting the cabling.

FDDI specifies multiple types of media: 62.5/125μ multimode fiber, 50/100μ single mode fiber, shielded twisted pair, and unshielded twisted pair type 3. It is recommended that the primary media for FDDI in the backbone be multimode fiber (MMF) or single mode fiber (SMF). Mixing of MMF and SMF in the same network is allowed so long as the maximum network span is within 200 km. If fiber is not installed, retrofitting existing wiring ducts to carry fiber is easier than retrofitting for copper cabling. Terminating the fiber for backbone applications is not so expensive because of the few connections needed. The user should select the secure room to which to run the fiber and terminate in a wall-plate and patch-panel. The fiber patch-panel should be maintained in a clean room. At least six pairs of fiber should be pulled in each cable with two single mode fiber pairs and four multimode fiber pairs. The fiber specifications are described in chapter six.

Another issue in architecting an FDDI backbone is the scalability and head-room in terms of bandwidth requirements and availability. For example, the backbone may currently connect four Ethernet networks together on one floor of building A and three token-ring networks on another floor of the same building. Also, the FDDI connects to another building B which has two Ethernet networks on the first floor. If the token-ring networks are 4 Mbps and the

Ethernets are 10 Mbps, the total raw bandwidth requirement is $60 + 12 = 72$ Mbps. Although this is less than the 100 Mbps that FDDI affords, adding another Ethernet or 16 Mbps Token-Ring loads the backbone to near maximum. This in itself may not be bad. Usually, the local networks are connected to the backbone via a router or filtering bridge. This eliminates unnecessary traffic on the backbone. Therefore, the actual traffic on the backbone will be $\alpha \times 72$ Mbps, where usually $0.1 = < \alpha \leq 0.6$. If the backbone traffic starts exceeding 50% of the FDDI bandwidth (sustained) or the number of nodes on FDDI starts exceeding 80 (some companies recommend up to a 100) then the FDDI itself should be segmented into two or more FDDI networks connected via FDDI-FDDI bridges, routers or brouters. It is also advisable to examine the topology of the network if the FDDI backbone load starts getting very high. For example, it is possible that a group of engineering workstations on Ethernet are continuously accessing a server across the FDDI on a different Ethernet. In this case, it may be advisable to connect the FDDI workstations and servers directly to FDDI. This reduces the traffic to the bridge or router, reduces the latency through the bridges and routers and improves the performance of the workstations and servers. If the number of hops across bridges and routers is an issue, each of the FDDI subnets could be connected to an ATM or FDDI switch.

As the bandwidth requirement at the desktop increases, plans should be in place to migrate FDDI to the desktop. The migration of FDDI to the desktop should be *evolutionary* and not *revolutionary*. This can be best illustrated by figure 10.4b. The high-performance computers (e.g. servers) requiring higher bandwidth can be migrated from Ethernet to FDDI leaving the rest of the sub-networks intact. The FDDI/Ethernet bridge is replaced with a concentrator requiring minimal cabling change. The concentrator connects the servers to the FDDI backbone leaving the rest of the endstations on the Ethernet. Eventually, all the desktops could be FDDI equipped using a *ring of concentrators with a tree of bridges, routers, servers and end-stations* architecture. This migration will have minimal impact on the FDDI backbone.

10.2.2 FDDI and Cabling for Existing Networks

Most commercial buildings today are wired for Ethernet, Token-Ring, AppleTalk or some other LAN. Each of these networks can have different types of cable (UTP category 2, 3, 4 or 5, STP, coaxial, or fiber), different connectors (RJ-11, RJ-45, BNC, or MIC) and different topologies (star, bus, ring). In a heterogeneous environment, this can be very problematic. With different data networks, there are different configuration requirements leading to differences in the type of patch-panel, types of patch-panel cords, and cross-connect (or punch-down block). It is not possible to have all the different types of cable to the desktop with all the different connectors. The user has to standardize on one or at most two media types that can be run to every desktop and the connectors to terminate the cables in.

If the user decides to have FDDI to the desktop and finds it too expensive to install fiber to the desktop, FDDI can also run over copper wiring. The ANSI X3T9.5 (FDDI) committee is currently developing a specification for FDDI over copper. This effort is called twisted pair FDDI or TP-FDDI. The user is cautioned that efforts to run FDDI over UTP type 3 (or voice-grade) cable have thus far been unsuccessful, and TP-FDDI has settled on STP and UTP Type 5 wire over 100 m for immediate standardization. It is interesting to note that although Ethernet is specified to work over UTP type 3 or voice grade UTP, many vendors recommend using the better quality UTP type 5 cable. Type 5 cable is not more expensive than type 3, and market surveys have shown that increasingly type 5 is being installed as the cable of choice rather than voice-grade. Type 5 cable is suitable for most LANs such as Ethernet, Token-Ring, ARCnet, and FDDI.

10.2.3 Designing an FDDI Desktop Network

If the desktop FDDI network is a small workgroup environment such as a CAD/CAM group or an engineering group, it may be cost-effective to connect the desktop to a stand-alone concentrator via copper wiring (as specified in the TP-FDDI recommendations). The copper cabling can be STP or UTP Type 5. This wiring may already exist for Token-Ring or Ethernet networks. If not, ad hoc wiring (figure 10.6a) may be implemented for the workgroup, although structured wiring (figure 10.6b) is recommended for scalability.

For the desktop connection, FDDI over fiber can be very expensive because very few buildings have fiber installed to each cubicle, and the cost of terminating fiber to the desktop is very high. If the user desires to use fiber because of security, fault-tolerance, or noise immunity reasons, at least two pairs should be laid to the desktop. Each pair should be terminated with the MIC connector as specified in the FDDI PMD specification. This type of connector is also used with other fiber-based LANs such as fiber-channel and ATM.

If the building is new or is being rewired, depending on the cost structure, one or both of the following should be implemented:

- A four-pair bundle of at least 100 m UTP Type 5 from patch-panel to desktop. This is recommended instead of the voice grade UTP because it allows higher-speed networks to be implemented without changing the cable plant.

- 62.5/125 μm MMF from the wiring closet to the desktop in addition to the UTP Type 5 connectors. These desktop connections should be ter-

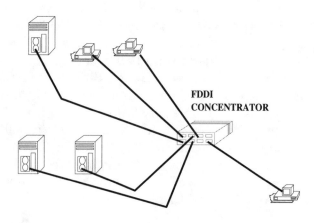

—————— STP or UTP Type 5 cabling

Figure 10.6a   An FDDI workgroup network with a stand alone FDDI concentrator.

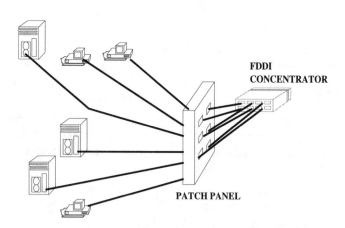

—————— STP or UTP Type 5 cabling

Figure 10.6b   An FDDI workgroup network with a stand alone FDDI concentrator from the patch-panel.

minated and connectorized (a MIC connector) as a wall outlet. If budget permits, at least two such connectors should be available in the cubicle. This makes the workplace cable ready for the next generation higher-speed networks such as ATM.

CHAPTER 11

Performance Characteristics and Tuning

11.1 FDDI CHARACTERISTICS

FDDI can be characterized by a number of different parameters. For the user, the parameters of importance are the maximum distance between stations, maximum span of the network, maximum number of stations, types of cables, topology, throughput, and latency. There are many pitfalls to installing a network, and unfamiliarity is the most dangerous one. Knowing the physical limitations of a network and the performance issues can help avoid embarrassing situations such as installing an FDDI network and discovering that it delivers Ethernet-like performances.

11.1.1 Maximum Distances and Number of Stations

FDDI was designed to support up to 1,000 physical connections and a total fiber length of 200 kilometers with 100 kilometers per ring. These values represent *defaults* which were used to calculate all default parameter values. There is one physical connection per SAS and two per DAS. A concentrator has two or more physical connections. Therefore the actual number of stations that can be supported is dependent on the type of station. In a DAS-only network, the maximum number of stations supported would be 500.

The maximum latency through a station is composed of:

- Latency through the PHY component(s)

- Latency through the MAC component(s)

The latency through a node is the time it takes for the starting delimiter of a frame to travel from the optical input of the MIC (where it enters the node) to the optical output of the MIC (where it leaves the node, possibly a different MIC). The range of the per port latency through a node is [240 ns, 1,164 ns][1]. The rule of thumb is to assume a 1,000 ns latency per port. The latency through the MAC is less than 240 ns, and in the repeat configuration is 80 ns.

The total ring latency is the sum of the latencies of the individual stations and the latency of the links. The latency of the links can be calculated

---

1. The PHY-2 specified number is 1,164 nanoseconds. The X3.146-1988 PHY specified number is 756 nanoseconds.

from the propagation delay of light through the fiber which is 5.085 microseconds/kilometer. The per port latency is approximately 1 microsecond. Therefore a 500 node DAS ring with 200 kilometers of cabling has a total latency of 1.773 (according to the original PHY standard).

In practice, it is not advisable to install an FDDI network reaching the maximum configuration for several reasons:

- Applications requiring low latency and high throughput may encounter significant access delays and reduced throughput depending upon the TTRT and network load.

- The network becomes much more sensitive to parameter adjustments such as TTRT, synchronous bandwidth allocation, and peak traffic patterns.

- Network stability and manageability decreases because of the increased number of entities.

It is recommended that the number of stations on an unsegmented FDDI be less than 80 nodes. Some vendors recommend not exceeding 100 nodes on the same network.

It is always possible to trade off the number of nodes versus the ring size. If a situation arises where the number of nodes is few but the inter-node distance exceeds the FDDI multimode fiber specification of 2 km, it is possible to use single mode fiber to increase the maximum inter-nodal distance from 2 to 40 km. The number of nodes that FDDI can support can be correspondingly increased by reducing the link lengths. This extension of the defaults may require the TVX and TRT values to be fine-tuned. A more subtle impact is on the link error threshold. If the threshold is lower than the FDDI default of 10^{-8} and the network latency is around the maximum tolerable, a small TVX value can cause the ring to reinitialize when a number of back-to-back frames are in error. The TVX will expire because a downstream station does not see valid frames for TVX time even though the token has not been lost.

In order to support more than 500 stations or to speed up the network recovery times, all timer-based parameter values would have to be recalculated. This may be difficult with some FDDI chipsets supporting the default maximum values. The user is cautioned against fine-tuning the ring recovery times by changing the TVX value because the benefit obtained is marginal and the risk of destabilizing the network is increased.

Normally, the TVX and TRT registers are not accessible to the systems programmer.

11.1.2 Efficiency

The efficiency of a network can be defined as the ratio of throughput to the bandwidth. The FDDI bandwidth is 100 Mbps. Its actual throughput under various loads is less than 100 Mbps. This is mainly due to token access delay caused by ring latency. The efficiency [1] of an FDDI network can be approximated by the equation:

$$(T-D) / T; \text{ where } T = \text{TTRT in ms, and } D = \text{ring latency in ms}$$

For a given TTRT, the larger the ring, the lower the efficiency. Conversely, the larger the TTRT, the higher the efficiency. Unfortunately, high TTRT adversely affects the network access delay when the network is operating near saturation or under extremely high burst rates.

Most networks do not operate near saturation. In fact, a recommended practice amongst MIS managers is to segment or upgrade a network when the load starts approaching 50%. Most networks will not be optimized for efficiency but will be optimized for throughput and latency.

11.1.3 Throughput

The throughput of a ring is the percentage of the cumulative offered load which is carried by the network. The throughput is equal to the offered load at light loads. That is, if the offered load on the ring is 30 Mbps, the throughput is also 30 Mbps. If the offered load on the ring is 400 Mbps, the throughput will be less than 100 Mbps. The maximum throughput under high load has also been referred to as usable bandwidth in the literature.

A network manager is interested in achieving a throughput close to the theoretical maximum throughput. The actual throughput is a function of the type of upper layer protocol, the class of service (asynchronous or synchronous), the TTRT value, and the ring size. Consider a large ring with a latency (D) of 1 ms with a TCP-large-window-size implementation as the upper layer protocol. This implementation of TCP allows large amounts of data to be transmitted before pausing for an acknowledgment. This implementation provides increased throughput on high-speed networks. If the TTRT (T) is set to 8 ms, the host has a very fast CPU and the TCP window size is set equal to 128 KBytes, for extended transfers, the station is restricted to transmitting for the 7 ms (T to D) because when the TRT expires the station has to release the token. This is assuming no station transmitted on the previous rotation and this station receives the token after one ring latency (1 ms). Following the 7 ms transmission, the station has to wait for the token to rotate twice around the ring (assuming it is the only station transmitting) before it can transmit again as the Late_Flag will have been set on the previous rotation. Therefore, to calculate the actual throughput:

It takes 10.24 ms to transmit 128 KBytes of data at 100 Mbps.

The actual time taken is:

7 ms on the first token rotation. The token is then released as TRT expires.

+ 2 ms (2 * ring latency). The token rotates twice around the ring as Late_Ct was set.

+ 3.24 ms. Time taken to transmit the remaining 40.5 MB of the TCP window

= 12.24 ms

The actual TCP throughput is (10.24 / 12.24)* 100 Mbps = 83.67 Mbps.

We can see that the application is throttled to 83.67 Mbps even though the network and the application are capable of exceeding that. With a 16 ms TTRT, the throughput can be improved by up to 10% to the near maximum theoretical throughput (about 93 Mbps).

11.1.4   Types of Network Delays

The third performance parameter of interest to the network manager is the maximum end-to-end delay of the network. This is the time since the first bit of a packet is queued for transmission to the time that the first bit is received by the receiver. There are several components to the end-to-end delay or latency:

Queuing delay is the most important parameter for latency critical applications in LANs. This is the time spent by the packet waiting in queue behind other packets. This is the time since the first bit of the packet is queued for transmission at the transmit queue by the application, to the time that the first bit is transmitted on to the network.

Transmission time is fixed by the network transmission rate and the size of the frame. In the case of FDDI, the transmission rate is 80 ns per byte which is a transmission time of 0.36 ms for a 4,500 byte packet.

Propagation delay is the time taken by the first bit of a packet to travel from the source to the destination. It is determined by the spatial location of the source and destination pair. Typical FDDI networks have a ring latency in the range of 100 microseconds. The maximum latency can be approximately 3.4 ms.

Receive time is the time between the arrival of the frame at the destination and the reception of the last bit. This is equal to the transmission time.

In a LAN, the queuing delay is the most critical component of the total delay. Slightly less critical is the queuing delay at intermediate stations such as bridges and routers. Propagation delay is negligible because of the small geographical distances.

For a latency critical application, it is desirable to provide a bound on the access delay and the lower the bound the better it is. The FDDI synchronous channel provides such a bound.

If n = number of stations, T = TTRT, and D = ring latency:

- The maximum access delay in a synchronous only network is approximately T + D.

- The maximum access delay in an asynchronous only network is (n − 1)*T + 2*D.

- In a mixed asynchronous and synchronous network, the maximum access delay for synchronous traffic is 2*T.

The maximum delay occurs with saturation loads (offered load equals network capacity) or under very bursty loads which is typically not the case. In most networks, the asynchronous service can provide adequate latency guarantees. However, it is better to use the synchronous channel for multimedia applications with video and audio components, and to use the asynchronous channel for normal data applications. The allocated synchronous bandwidth can always be used by the asynchronous transmissions when it is not being used. Thus, no bandwidth is wasted.

11.1.5   Ring Latency

The effect of increasing or decreasing the ring latency is less noticeable than the effect of increasing or decreasing the load on the network. Although the mean, variance and maximum access delay increase when the ring size increases, the effect is less noticeable for the synchronous transmissions. In fact, in a recent study of the impact of TTRT [2], ring size and various traffic types on the transmission of multimedia over FDDI, the conclusion was that there is very little impact on the maximum access time for 99% of all packets transmitted over the synchronous channel. However, increasing the ring latency without increasing the TTRT reduces the efficiency.

11.2   SETTING THE PARAMETERS

The only parameter which the user can change to optimize data movement is the TTRT. No other parameter should need to be changed for most topologies.

A rule of thumb is to maximize throughput if the network is data only. In this case, latency is not an important issue and the TTRT values should be set high to obtain maximum throughputs. A TTRT in the range of 16 to 40 ms is appropriate for such applications.

The greater the number of stations, the larger the TTRT should be.

If the network also supports very bursty traffic such as imaging, the TTRT should be set to 20 to 60 ms. The higher TTRT values ensure that image bursts do not encounter significant queuing delay at any station.

The larger the size of image files, the larger the TTRT should be.

If it is an integrated voice, video, data and imaging network, the TTRT should be set to approximately 16 ms for a network of up to 80 nodes. This TTRT value offers high throughput for the data and imaging applications and low latency for voice, video applications.

The smaller the TTRT, the lower the latency.

It should be noted that TTRT values below 8 ms can decrease throughput and efficiency.

For other configurations such as extremely large networks (greater than 200 nodes), it is still possible to obtain low latency and high throughput but the TTRT value depends on the application characteristics.

It should be noted that the TTRT is the lowest winning T_Req. Hence if a station desired a 40 ms TTRT and another station bid 16 ms, the 16 ms T_Req will become the TTRT. If the user finds that the TTRT is not within the suitable range, the user should change the T_Req of the offending station (16 ms bidder). Often the TTRT is not available to the systems programming interface. In that case, the user should request the FDDI vendor to change the T_Req or provide the user with the systems programming interface to change it to the adequate value.

11.3 BIBLIOGRAPHY

1. R. Jain, "Performance Analysis of FDDI Token Ring Networks: Effect of Parameters and Guidelines for Setting TTRT". ACM SIGCOMM'90 Symposium on Communications Architectures and Protocols, Sept. 1990.

2. A. Shah, I. Rubin, et. al, "Multimedia Over FDDI". IEEE Local Computer Networks Conference, Minneapolis, Sept. 1992.

CHAPTER 12

FDDI Design and Troubleshooting

In this chapter, issues in designing hardware and software for an FDDI network adapter are discussed. Both non-intelligent and intelligent versions of network adapters are considered. This is followed by a discussion on troubleshooting an FDDI network.

12.1 HARDWARE DESIGN

Network adapters are used in an end station for data, video and voice transfers between different stations in a network. The different applications running over a network include file transfer, virtual terminal emulation, electronic mail, multi-media etc. Network adapters can be divided into two categories:

- Non-intelligent adapters
- Intelligent adapters

While a given application can be run using either a non-intelligent or an intelligent network adapter, an intelligent adapter provides better performance compared to a non-intelligent adapter. A network adapter consists of three functional blocks (as shown in figure 12.1):

- Network interface
- System interface
- Bus interface

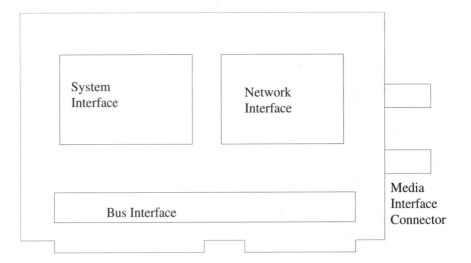

Figure 12.1   Block diagram of a network controller.

The network interface implements the hardware and software specific to the network to which the adapter is connected. For instance, in an FDDI network adapter, the network interface part consists of PMD, PHY, MAC and SMT related hardware and software components. The system interface deals with buffering and structuring the packets that are to be transmitted and received. This block also deals with the technique used for transferring data from the network adapter to the host system and vice versa. Bus interface deals with hardware related to a specific bus architecture for which the adapter is designed (such as data bus width, bus mastering capabilities etc.).

12.1.1   Non-Intelligent FDDI Network Adapter

In a non-intelligent adapter, the host system controls all the devices in the adapter. Figure 12.2 shows the block diagram of a non-intelligent FDDI adapter. Local buffer memory is required if the network receives data at a rate exceeding that of the transfer rate from network controller to system memory. The size of the local buffer memory is determined by the system bus latency and the bus interface logic. In some designs, the system interface may have some amount of FIFO, and in such cases local buffer memory may not be required. In systems employing non-intelligent adapters, the SMT is implemented in the host. Non-intelligent adapters are typically used in low performance and low-cost systems such as PCs. The disadvantages include lower performance, and overhead on host processor and device drivers.

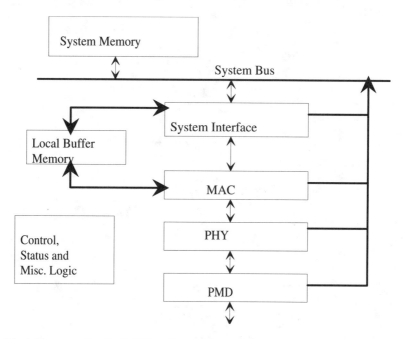

Figure 12.2   Block diagram of a simple FDDI (SAS) adapter.

12.1.2 Intelligent FDDI Network Adapter

In an intelligent FDDI adapter, some kind of processing power resides on board. This processing power may be in the form of a microcontroller, CISC, or RISC processor. Figure 12.3 illustrates the block diagram of a typical intelligent FDDI adapter.   The range of functions that need to be performed on board determines the kind of processing power. For instance, if the function of the on board processor is only to perform SMT, a low-end processor such as 80186 is sufficient. But if the on board processor is expected to perform multiple functions such as SMT and upper layer protocol processing, a more powerful processor such as Am29000 and i960 is required. The NP memory is a combination of volatile and non-volatile memory. Non-volatile memory is used for storing the executable part of the code. Volatile memory may be used for storage requirements such as management information base (MIB), operating system stack, and transmit and receive buffers. Typically, intelligent adapters also implement a content addressable memory (CAM) in order to detect multicast and alias addresses.

12.2 SOFTWARE DESIGN

Software for a FDDI (intelligent or non-intelligent) adapter consists of two components:
- A standard network driver such as Packet Driver, NDIS, ODI, or STREAMS
- SMT

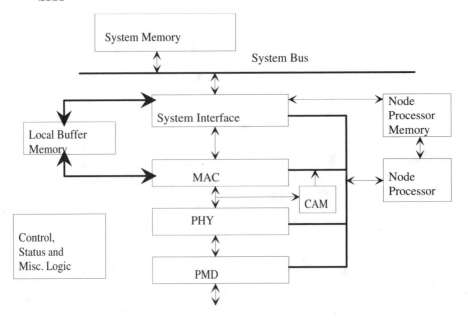

Figure 12.3 Block diagram of a sophisticated FDDI (SAS) adapter.

To provide standard interfaces to network operating systems (NOS), several NOS vendors have developed specifications which make the network hardware independent of the NOS. Novell has the Open Data link Interface (ODI) specification, Microsoft/3Com has the Network Device Independent Specification (NDIS) and UNIX has STREAMS. Thus, adapters from different manufacturer's provide the same interface to the NOS if the device driver is written to the device independent driver specifications. An adapter is guaranteed to work with the standard NOSes if the driver software has been written to one of the above NOS driver specifications. In these standard device drivers, there are two components:

- Hardware dependent (which is provided by the adapter vendor).

- NOS dependent (the adapter vendor uses the standard interfaces provided by the NOS).

SMT can be implemented as an application, as an extension to the device driver or as firmware in the adapter. If the network adapter is non-intelligent, SMT is implemented in the host system. In such cases, SMT is typically implemented as a terminate and stay resident (TSR) module in a non multi-tasking environments (such as DOS), and as a process in multi-tasking environments (such as UNIX). In Novell Netware environments, SMT can be implemented as a Network Loadable Module (NLM). If the adapter is intelligent, SMT is implemented on the adapter and this minimizes the SMT traffic over the system bus. The other advantage with this approach of implementing

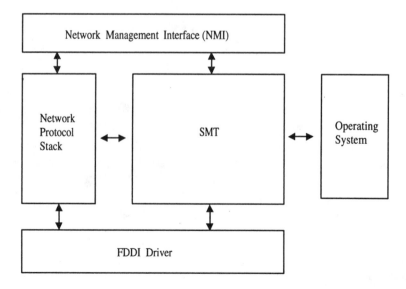

Figure 12.4 SMT interfaces.

SMT on the adapter is that SMT is transparent to the host system. So porting of driver software to adapters with different FDDI chipsets becomes easier. The disadvantages are:

- Higher cost

- The system interface logic needs to be more sophisticated (since the SMT frames need to be identified and directed to the node processor memory in the adapter).

SMT is typically designed as an independent module with well defined interfaces for different entities such as the operating system, buffer memory, FDDI chip specific drivers, and management application process (figure 12.4). These well defined interfaces allow the SMT be ported to different software and hardware environments. The interface for operating system provides services such as creating and deleting a *process* and also changing the priority of a process. In cases where the operating systems cannot provide a process creation facility, SMT contains a lightweight scheduler. The buffer memory interface provides functions for allocating memory space when SMT needs to receive a frame or build a frame for transmission. It also provides a dynamic memory allocation function so that SMT can use memory when it is required. The FDDI chip driver interface contains routines that manipulate the network related hardware. For instance, a function for transmitting a given line state or a frame is provided by this interface. Other functions provided by this interface include indication of a given line state to SMT and detection of LEM events. The functions provided by the management application interface include:

- Access the MIB attributes for read/write operations

- Start and stop the SMT

- Make requests to SMT to transmit and receive SMT frames (such as ECHO, PMF request/response frames).

12.3 TROUBLESHOOTING AN FDDI NETWORK

Troubleshooting is the most interesting and probably the most difficult process of bringing up a network. This is especially true in the case of a new network. There are two types of FDDI network analyzers:

- Passive monitors

- Active monitor

Passive monitors captures data on the ring by tapping 10 to 15% of the optical signal on the media. Such monitors do not repeat or transmit data on the network. Active monitors on the other hand form a part of the ring and they receive, repeat and transmit data on the network. Active monitors are useful since one can query any node in the network by sending a request to that node. Active monitors can, if desired, modify the data on the ring (e.g. the A, C and E indicators may be set by a active monitor). There are other low end monitoring equipment which can be used to monitor the optical signal levels and perform limited physical layer testing.

12.3.1   Troubleshooting an FDDI Dual Ring

A dual ring with four dual attachment stations and an active FDDI network analyzer (figure 12.5) will be the basis for the following discussion. Figures 12.6 through 12.13 explain the step-by-step procedure for troubleshooting a FDDI dual ring.

Figure 12.6 identifies a problem in the configuration. If the configuration is correct, figures 12.7, 12.10 and 12.11 can be used to determine any problem with the path associated with the SMT and LLC traffic. In the case of a configuration problem, figures 12.9 and 12.12 are useful in identifying and troubleshooting a misbehaving node. Once the basic problems are identified and solved, figure 12.8 can be used to determine the long term stability of a ring.

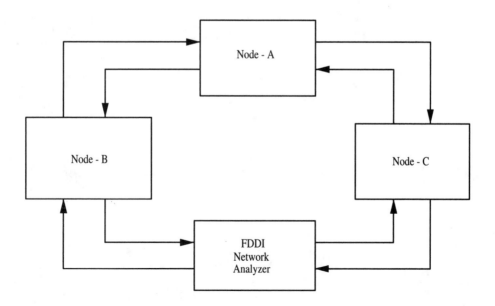

Figure 12.5   Troubleshooting an FDDI dual ring.

12.3.2   Troubleshooting an FDDI Tree Structure

A tree structure with a DAC as a root and three SAS nodes (as shown in figure 12.14) is considered. A simple step-by-step procedure for troubleshooting is given in figures 12.15 to 12.19. Figures 12.15 and 12.16 are used to identify a configuration problem. Figures 12.17 and 12.18 identify any problem with the path associated with SMT and LLC traffic. Stability of a ring is determined by using figure 12.19.

In the above discussions, simple and more common configurations, such as a single MAC and dual path DAS and DACs are considered. However, FDDI allows for more complex configurations (i.e. multiple MACs and paths). In such cases, the troubleshooting becomes more difficult. Care should be taken in connecting the nodes. Not using standard MIC connectors and proper keying can lead to multiple problems (such as twisted rings and M to M port connections).

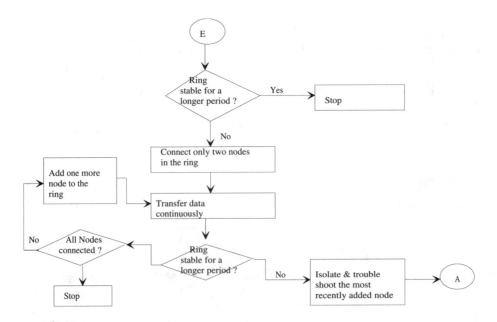

Figure 12.6   Troubleshooting an FDDI dual ring.

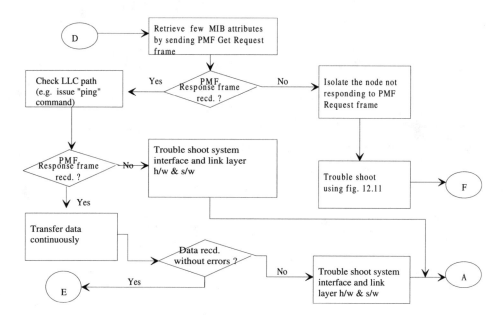

Figure 12.7   Troubleshooting an FDDI dual ring.

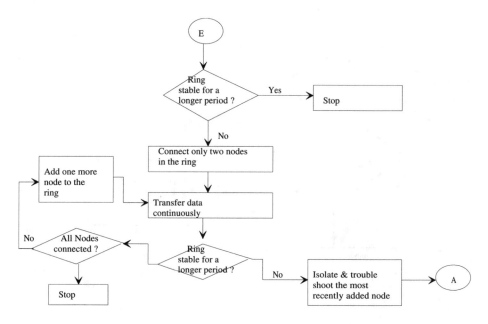

Figure 12.8   Troubleshooting an FDDI dual ring.

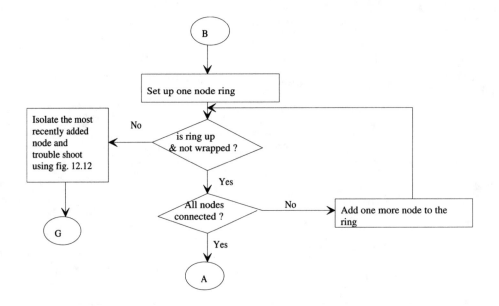

Figure 12.9   Troubleshooting an FDDI dual ring.

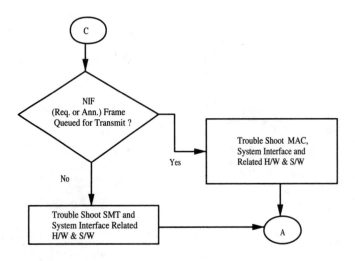

Figure 12.10   Troubleshooting an FDDI dual ring.

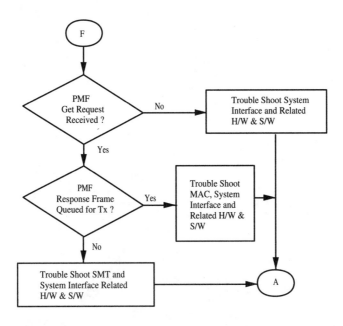

Figure 12.11   Troubleshooting an FDDI dual ring.

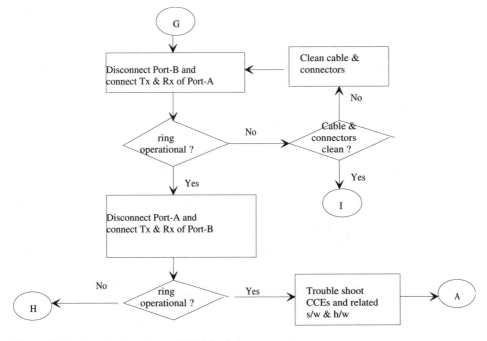

Figure 12.12   Troubleshooting an FDDI dual ring.

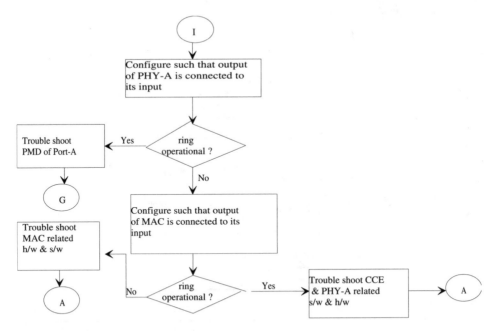

Figure 12.13   Troubleshooting an FDDI dual ring.

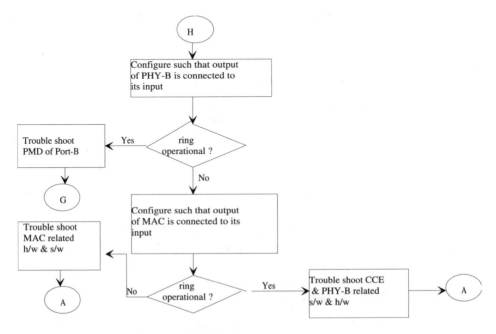

Figure 12.13a   Troubleshooting an FDDI dual ring.

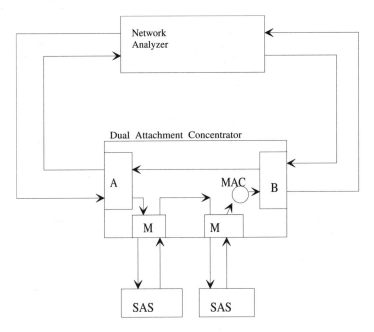

Figure 12.14   Troubleshooting an FDDI tree.

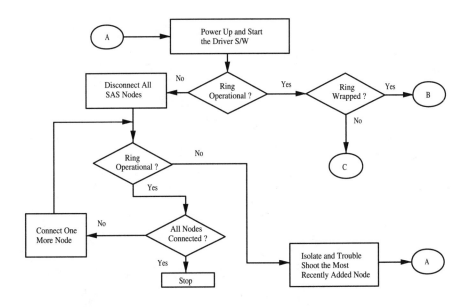

Figure 12.15   Troubleshooting an FDDI tree.

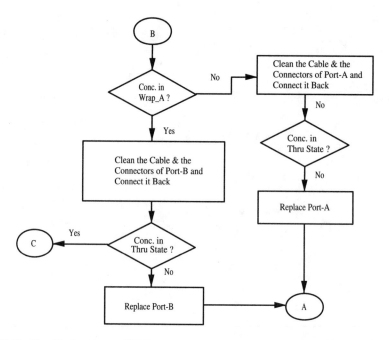

Figure 12.16   Troubleshooting an FDDI tree.

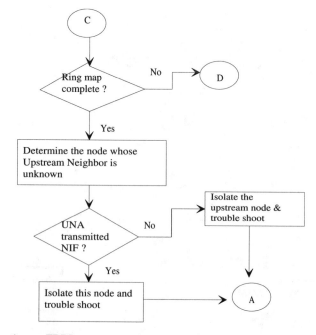

Figure 12.17   Troubleshooting an FDDI tree.

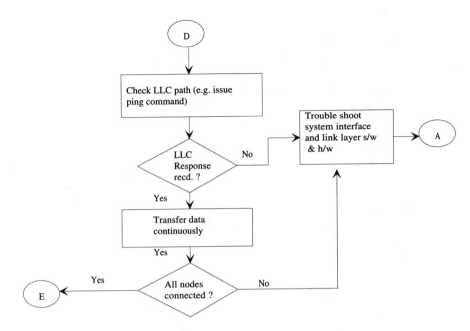

Figure 12.18 Troubleshooting an FDDI tree.

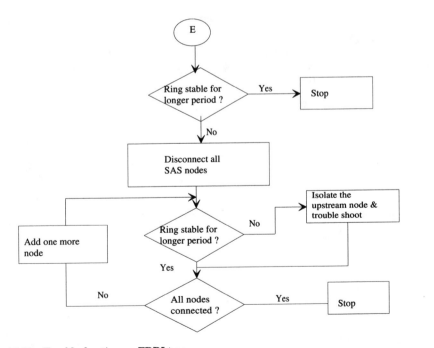

Figure 12.19 Troubleshooting an FDDI tree.

Glossary of Terms

A_Max	Maximum time to acquire the signal
ALS	Active Line State
ANSI	American National Standards Institute
APD	Avalanche Photo Diode. An optical detector with very high sensitivity.
ARP	Address Resolution Protocol
ASN.1	Abstract Syntax Notation One
ATM	Asynchronous Transfer Mode. A type of cell-relay network which uses fixed size cells (53 bytes).
B-ISDN	Broadband Integrated Services Data Network (CCITT standard)
BER	Basic Encoding Rules
CEM	Configuration Element Management
CFM	Configuration Management
CLNP	Connection Less Network Protocol
CMIP	Common Management Interface Protocol
CMT	Connection Management
Control Indicators	Indicators used for identifying the state of E, A, and C bits of the Frame Status field of a frame
CRC	Cyclic Redundancy Check
CSMA/CD	Carrier Sense Multiple Access / Collision Detect
D_Max	Maximum Ring Latency
DA	Destination Address
DAC	Dual Attachment Concentrator
DAS	Dual Attachment Station
DCD	Duty Cycle Distortion
DDJ	Data Dependent Jitter
DLL	Data Link Layer

DNA	The abbreviation for Downstream Neighbor Adress. FDDI is a logical ring and hence every station has an upstream neighbor and downstream neighbor. This is also a MIB attribute transmitted in a SMT Neighborhood Information Frame.
EB	Elasticity Buffer
ECF	ECho Frame. A type of SMT frame similar in function to the internet ICMP *ping*.
ECM	Entity Coordination Management
ECMA	European Computer Manufacturers Association - a standards body; pronounced *eck'ma*
ED	Ending Delimiter. A special symbol (T) used to demarcate the end of a frame
EIA	Electronic Industries Association
Elasticity Buffer	Facility in PHY to compensate for the difference between the local and the recovered clock
Error_Ct	A count of all error frames detected by this station and not by any previous stations. Frames received with the E indicator set are not counted as error frames by a station as they will have been counted at previous stations.
ESCON	Enterprise System CONnection
ESF	Extended Service Frame
F_Max	Maximum Frame Transmission Time (including 16 symbols of preamble)
FBMT	Frame Based ManagemenT
FC	Frame Control field
FCC	Federal Communications Commission – a government agency; often used in reference to regulations for RFI *FCC Regulations Part 15, Subpart J.*
FCS	Frame Check Sequence
FDDI	Fiber Distributed Data Interface. A 100 Mbps dual-ring network.
Frame_Ct	A count of all frames received by a station

FS	Frame Status. A 3 symbol field at the end of a frame (after the ending delimiter) which indicates the status of the frame. The status symbols are Error (E), Address recognized (A), and Copied frame (C)
FSM	Finite State Machine
GDMO	Guidelines for the Development of Managed Objects
GOSIP	Government Open Systems Interconnection Profile; pronounced *gossip*
HLS	Halt Line State. A state entered by a PHY when 16 or 17 consecutive Halt symbols are received.
I_Max	Maximum Station Physical Insertion Time
ICC	Intermediate Cross Connect
ICMP	Internet Control Message Protocol
IEEE	Institute of Electrical and Electronics Engineers
IETF	Internet Engineering Task Force. A de facto standards body famous for its TCP/IP and SNMP protocols.
ILS	Idle Line State. A state entered by a PHY upon receiving 4 or 5 consecutive Idle symbols. During connection management 16 or 17 consecutive Idle symbols have to be received to be recognized as ILS. It is often called *super ILS*
IP	Internet Protocol
IS-IS	Intermediate System to Intermediate System routing protocol. This is an ISO defined protocol to route in an OSI environment
ISO	International Standards Organization
L_Max	Maximum Transmitter Frame set-up Time
LAN	Local Area Network
Late_Ct	A count of the number of times a token is late at a station, arriving after the TRT has expired
LCF	Low Cost Fiber. An ANSI X3T9.5 standard which refers to the reduced cost of transceivers for multimode fiber with reduced distance-bandwidth requirements.

LEED	Link Error Event Detector. This is a function in the Connection Management part of SMT to identify error events. It is implementation dependent and is usually implemented in the PHY silicon.
LEM	Link Error Monitor. A mechanism for dynamic monitoring of link quality when the ring is operational.
LLC	Logical Link Control. The part of the Data Link layer which ensures an error-free link to link transport in the IEEE 802 LANs and FDDI.
Lost_Ct	A count of the frames that contain an invalid symbol other than I
LSU	Line State Unknown. State entered by a PHY when the criteria for exiting the current line state is met and the criteria for entering any other line state is not met.
M_Max	Maximum Number of MAC entities
MAC	Media Access Control. A sub layer of the Data link layer common in shared media LANs.
MAN	Metropolitan Area Network. A term used to refer to networks larger than LANs but smaller than WANs and typically spanning a distance of 100-200 miles, the size of a city.
MCC	Main Cross Connect
MFM	Modified Frequency Modulation. One of many bit encoding schemes. Used in some earlier hard drives.
MIB	Management Information Base
MIC	Media Interface Connector
MLS	Master Line State
MLT-3	Multi Level Transmission -3. A three level encoded transmission scheme used for CDDI.
MMF	Multi Mode Fiber as opposed to SMF. A type of fiber in which light travels along multiple paths. This is used in short-haul communications such as LANs.
MTU	Maximum Transmission Unit
NAC	Null Attachment Concentrator

NIF	Neighborhood Information Frame. A type of SMT frame.
NLS	Noise Line State. PHY enters this line state when 15 or 16 consecutive potential noise events occur.
Node	The generic name for any FDDI device such as repeater, concentrator, Dual Attachment Station (DAS), and Single Attachment Station (SAS). It is different from an FDDI station in that a station is required to have a MAC whereas a node may or may not have a MAC. For example a MAC-less concentrator is a node and a concentrator with a MAC is a station. Also see Station.
NRZ	Non Return to Zero. A type of bit encoding scheme.
NRZI	Non Return to Zero Invert. A type of bit encoding scheme used in FDDI.
NSA	Next Station Addressing; An SMT frame type.
OA&M	Operations, Administration & Maintenance (telephone industry jargon)
OSI	Open Systems Interconnection
OSPF	Open Shortest Path First routing protocol. This is an Internet Engineering Task Force (IETF) defined routing protocol for routing in TCP/IP networks and is specified in RFC 1247.
PA	Preamble. This is eight bytes of idle symbols transmitted between frames (including a token). This allows the PHY Elasticity Buffer to recenter and the MAC to distinguish between back-to-back frames. The preamble also serves the additional purpose of providing clock.
PCM	Physical Connection Management
PDU	Protocol Data Unit. The unit of transfer of information from protocol to protocol. Therefore TCP PDUs, IP PDUs and LLC PDUs.
PHY	PHYsical layer
PIN	Positive Intrinsic Negative diode. A commonly used optical detector.

PMD   Physical Media Dependent layer

PMF   Parameter Management Frame. An SMT frame used
                     for accessing MIB attributes.

Preamble   Preceding every frame eight bytes of Idle symbols are
                     transmitted. This is called the preamble or Inter-
                     Frame-Gap (IFG). This is used to delineate the
                     individual frames and for clock recovery purposes.

QLS   Quiet Line State. This is the line state where
                     minimum or zero power is transmitted. Detection of
                     QLS indicates either the link is broken or the
                     connection is being restarted.

RAF   Resource Allocation Frame. This frame can be used by
                     the bandwidth allocator to implement the synchronous
                     bandwidth allocation protocol.

RDF   Request Denied Frame. This is an SMT frame used to
                     indicate a problem with the SMT Request frame. It is
                     generated in response to individually addressed
                     request frames and carries a reason code for the
                     denial.

Repeat Filter   This function is implemented on the transmit path in
                     the PHY to prevent propagation of code violations and
                     invalid line states. This function is normally carried
                     out by a MAC but on some internal paths in a
                     concentrator there is no MAC in the path and the
                     repeat filter becomes necessary.

RIP   Routing Information Protocol. This is an Internet
                     Engineering Task Force (IETF) defined routing
                     protocol for routing in TCP/IP networks popular in the
                     older BSD 4.2 UNIX implementations. It is specified
                     in . This is now being replaced with the OSPF
                     protocol.

RMT   Ring ManagemenT. A module of the SMT specification
                     which manages a MAC in a station. There is one RMT
                     per MAC.

SA   Source Address. This is a 16-bit or 48-bit field. If the
                     most significant bit is one it indicates a source-routed
                     frame.

SAC	Single Attachment Concentrator. An FDDI node which distributes a ring in the tree. It is not at the root of the tree. It attaches up the tree via an S-port and typically connects to an M-port of another concentrator.
SAS	Single Attachment Station. An FDDI station which connects to a single ring. It connects to a concentrator which distributes a single ring to multiple single attachment stations.
Scrambler	Logic implemented to distribute the energy of the transmission over lower frequencies in order to avoid peaks which can cause electro-magnetic interference and cross-talk. This function is implemented in transmissions over copper cables.
Scrubbing	Mechanism for removal of orphaned frames, errored frames, and fragments.
SD	Start Delimiter. This is the frame beginning delimiter. The symbols J (11000) and K (10001) together constitute the start delimiter.
SDU	Service Data Unit
SIF	Station Information Frame. An SMT frame based management frame type. SIF frames can be of type operation or configuration. It is used to carry information about the station such as type of station, number of ports, MAC neighbors, etc.
SMAP	System Management Application Process
SMF	Single Mode Fiber as opposed to MMF. A type of fiber in which light travels along one path only. Used in long distance communications.
SMI	Structure of Management Information
Smoother	A state machine in the PHY which inserts or removes upto 3 bytes of idle symbols in the inter-frame gap. It inserts idles if the inter-frame gap is less than 14 symbols and removes symbols if the inter-frame gap exceeds 14 symbols.
SMT	Station ManagemenT. The fourth of the FDDI set of standards defining the management protocol for the

	FDDI MAC, PHY and PMD. It defines node, link and remote ring management features.
SNMP	Simple Network Management Protocol. A popular network management protocol defined by the Internet Engineering Task Force. It is used to perform remote management of various networks.
SRF	Status Report Frame. A frame generated by a trap or interrupt-driven Status Report protocol which is triggered by various events and conditions. It is similar to the trap feature of SNMP.
Station	The generic for any FDDI device which has a MAC. For example Dual Attachment Stations (DAS), Single Attachment Station (SAS) and a single-MAC Dual Attachment Concentrator.
STP	Shielded Twisted Pair. A type of copper cable which has better noise immunity and lower noise susceptibility than the regular phone cable (UTP Type 3). Commonly used in Token Ring networks.
Stripping	A mechanism employed by a station MAC to remove frames it transmitted
Symbol	FDDI transmits encoded data. A nibble (four bits) of data is encoded into a symbol (five bits) using 4B/5B encoding. Four bits can represent sixteen data numbers. Thus sixteen 5B combinations (out of a possible thirty-two) are selected to represent the numbers 0-15 for transmission.
T_Bid	Requested Token Rotation Time carried in a Claim frame (a 32-bit twos complement value)
T_Neg	During the claim process, before the token is issued or received the intermediate negotiated token rotation time is maintained in a T_Neg register. Once the claim process completes, the T_Neg is loaded into the TTRT register.
T_Opr	The operational value of the TTRT
T_Req	Requested Token Rotation Time (a 32-bit twos complement value). This is the requested TRT which is carried as the T_Bid value in the information field of

a claim frame during the claim process. It is a programmable value loaded by the (software) driver into the FDDI MAC. This value is used in the Claim process to decide how fast the token should rotate around the ring (on an average). T_Req is often not available to the systems programmer and a default value is programmed by the FDDI equipment vendor. The normal default value is 165 ms. Other commonly used values are 40 ms, 24 ms, and 16 ms.

TCC   TeleCommunications Closet.

TCP   Transmission Control Protocol. A popular connection-oriented transport protocol implemented over the Internet Protocol and defined by the Internet Engineering Task Force.

THT   Token Holding Timer. A timer which contains the time for asynchronous transmissions. It is loaded with the TRT value at the time the token returns. It is not enabled during synchronous transmissions.

Token   A special frame in FDDI used to control access to the media (fiber or copper). The token consists of Start Delimiter, Frame Control, Ending Delimiter: JK 80 TT. A station has to capture the token in order to transmit. At any given time only one station can transmit. After completing its transmission a station releases (issues) the token which can then be captured by a downstream station desiring to transmit.

TRT   Token Rotation Timer. A timer which times the actual rotation time of the token with respect to the target token rotation time. The difference between the target and actual is used for asynchronous transmissions. It is reset on receipt of an early token.

TTR   Timed Token Rotation protocol. The basic FDDI MAC protocol based on the work of Robert Grow et.al.

TTRT   Target Token Rotation Time. The negotiated and accepted token rotation time which is used to load the TRT in every station. The default value is T_Max.

TVX   Timer Valid Xmisssions. This timer measures
the time between valid frame receptions.
Its expiration indicates a lost token.

ULS   Unknown Line State. See LSU.

UNA   The abbreviation for Upstream Neighbor Adress.
FDDI is a logical ring and hence every station has an
upstream neighbor and downstream neighbor. This is
also a MIB attribute transmitted in a SMT
Neighborhood Information Frame.

UTP   Unshielded Twisted Pair. A type of copper cable which
is widely used in telephony and Ethernet networks.
There are several categories of this cable of which the
most common is the DIW Type 3. A better version of
this cable is the Type 4 cable.

WAN   Wide Area Network. A common acronym for the long-
distance networks.

Index